Study Gui for

Whitney and Rolfes's

Understanding Nutrition

Tenth Edition

Lori W. Turner, Ph.D.
University of Arkansas

Australia • Canada • Mexico • Singapore • Spain • United Kingdom • United States

COPYRIGHT © 2005 Wadsworth, a division of
Thomson Learning, Inc. Thomson Learning™ is a
trademark used herein under license.

ALL RIGHTS RESERVED. No part of this work
covered by the copyright hereon may be reproduced
or used in any form or by any means—graphic,
electronic, or mechanical, including but not limited to
photocopying, recording, taping, Web distribution,
information networks, or information storage and
retrieval systems—without the written permission of
the publisher.

Printed in the United States of America
   2 3 4 5 6 7   08 07 06 05

Printer: Darby Printing Company

ISBN: 0-534-62231-3

For more information about our products,
contact us at:
**Thomson Learning Academic Resource Center
1-800-423-0563**

For permission to use material from this text or
product, submit a request online at
http://www.thomsonrights.com.
Any additional questions about permissions can be
submitted by email to **thomsonrights@thomson.com**.

Thomson Wadsworth
10 Davis Drive
Belmont, CA 94002-3098
USA

Asia
Thomson Learning
5 Shenton Way #01-01
UIC Building
Singapore 068808

Australia/New Zealand
Thomson Learning
102 Dodds Street
Southbank, Victoria 3006
Australia

Canada
Nelson
1120 Birchmount Road
Toronto, Ontario M1K 5G4
Canada

Europe/Middle East/South Africa
Thomson Learning
High Holborn House
50/51 Bedford Row
London WC1R 4LR
United Kingdom

Latin America
Thomson Learning
Seneca, 53
Colonia Polanco
11560 Mexico D.F.
Mexico

Spain/Portugal
Paraninfo
Calle/Magallanes, 25
28015 Madrid, Spain

TABLE OF CONTENTS

Preface iv

Chapter 1 — An Overview of Nutrition 1

Chapter 2 — Planning a Healthy Diet 15

Chapter 3 — Digestion, Absorption, and Transport 27

Chapter 4 — The Carbohydrates: Sugars, Starches, and Fibers 43

Chapter 5 — The Lipids: Triglycerides, Phospholipids, and Sterols 57

Chapter 6 — Protein: Amino Acids 71

Chapter 7 — Metabolism: Transformations and Interactions 85

Chapter 8 — Energy Balance and Body Composition 97

Chapter 9 — Weight Management: Overweight and Underweight 107

Chapter 10 — The Water-Soluble Vitamins: B Vitamins and Vitamin C 117

Chapter 11 — The Fat-Soluble Vitamins: A, D, E, and K 129

Chapter 12 — Water and the Major Minerals 141

Chapter 13 — The Trace Minerals 155

Chapter 14 — Fitness: Physical Activity, Nutrients, and Body Adaptations 169

Chapter 15 — Life Cycle Nutrition: Pregnancy and Lactation 181

Chapter 16 — Life Cycle Nutrition: Infancy, Childhood, and Adolescence 193

Chapter 17 — Life Cycle Nutrition: Adulthood and the Later Years 207

Chapter 18 — Diet and Health 217

Chapter 19 — Consumer Concerns about Foods and Water 231

Chapter 20 — Hunger and the Global Environment 243

Note: Your opinion as a customer is valuable to us at Wadsworth. If you would like to share your comments on this study guide, please write to: Wadsworth, Attention: assistant editor, 2220 Parklake Drive Suite 300, Atlanta, GA 30345.

PREFACE

Welcome to the fascinating world of nutrition. This guide has been designed to complement the text and facilitate learning. The three-hole punched and perforated pages allow you to insert, delete, and rearrange pages to suit your needs. The guide includes for each chapter a chapter outline, a fill-in-the-blank summary, a chapter glossary, chapter study questions, short-answer questions, a crossword puzzle (new to this edition), and sample test questions. In addition, many of the chapters have problems to solve or chemical structure identification exercises.

- **Chapter Outline.** The outline parallels the chapter headings and subheadings and provides space for your own notes. We suggest you read the chapter and use the outline to record key points. Remember to include any additional information the instructor may have provided.
- **Summing Up.** This section provides a brief summary of the chapter using a fill-in-the-blank format. After reading and outlining the chapter, read this summary and fill in the blanks.
- **Chapter Glossary.** This section provides terms and their definitions from the chapter. Understanding the terms is essential for understanding the text material and responding correctly to test questions. It is recommended that you memorize all of the terms prior to reading the text. Comprehension of the terms is key in successful completion of the course.
- **Assignments.** The assignments ask you to answer chapter study questions, complete short-answer questions, identify chemical structures and reactions, and solve problems. Each of these tasks is designed to reinforce concepts presented in the chapter. Chapter study questions allow you to review key concepts and prepare you to respond to essay questions on examinations. The crossword puzzles challenge you to recall key terms from each chapter.
- **Sample Test Questions.** This guide provides test questions that are similar to questions instructors may select for an examination. For example:
  - This guide may ask you to name the thiamin deficiency disease; your instructor may ask which vitamin deficiency causes beriberi.
  - This guide may ask you to define primary deficiency; your instructor may ask you to define secondary deficiency.
  - This guide may ask you to calculate the number of kcalories in a hamburger; your instructor may ask you to calculate the number of kcalories in a bowl of cereal.

  Students who only learn the answers to the sample study questions in this guide are likely to perform at a lower level than students who use these questions to learn concepts that may be tested. So when a question asks what the best food source of vitamin C is, remember that you should also be able to name food sources of other vitamins as well.
- **Answers.** Answers are provided for summing up, chapter study questions, assignments, and sample test questions. Check your answers and restudy areas that you are unsure of or missed. When checking fill-in-the-blanks, be aware that more than one word may satisfactorily complete a sentence. For example, "activity" or "exercise" may be equally correct in a sentence. When your answer differs from the one provided, consider whether you have used an acceptable synonym or whether you have missed the concept. When checking answers to the chapter study questions, keep in mind that only key points have been provided. To adequately answer this type of discussion question requires not only a recall of key points but also descriptions of interrelationships and explanations of how and why. A well-written answer synthesizes knowledge gained from previous lessons, related courses, and personal experience.

CHAPTER 1
AN OVERVIEW OF NUTRITION

CHAPTER OUTLINE

I. Food Choices

II. The Nutrients
    A. Nutrients in Foods and in the Body

    B. The Energy-Yielding Nutrients

    C. The Vitamins

    D. The Minerals

    E. Water

III. The Science of Nutrition
    A. Nutrition Research

    B. Research Versus Rumors

IV. Dietary Reference Intakes
   A. Establishing Nutrient Recommendations

   B. Establishing Energy Recommendations

   C. Using Nutrient Recommendations

   D. Comparing Nutrient Recommendations

V. Nutrition Assessment
   A. Nutrition Assessment of Individuals

   B. Nutrition Assessment of Populations

VI. Diet and Health
   A. Chronic Diseases

   B. Risk Factors for Chronic Diseases

Highlight: Nutrition Information and Misinformation—On the Net and in the News

SUMMING UP

People eat for many reasons other than to obtain the (1)_____ they need. Reasons people select the foods they do include: personal preference, habit, ethnic heritage or tradition, (2)_____ _____, availability, convenience, economy, positive and negative associations, (3)_____ _____, values, image, and nutrition.

The human body is made almost entirely of (4)_____ derived from food. The six classes of nutrients are: carbohydrate, (5)_____ (lipid), protein, (6)_____ (fat-soluble and water-soluble), minerals, and (7)_____. The first four are (8)_____; the first three provide (9)_____ in a form the body can use. (Alcohol also releases energy, but is not a (10)_____; it isn't used for (11)_____, maintenance, or repair of body tissues.) The (12)_____ nutrients must be obtained preformed from food. Food energy is measured in (13)_____, units of heat, or kilojoules, units of (14)_____.

The DRI—(15)_____ Reference Intakes—represent a set of values for the dietary nutrient intakes of healthy people. The four values include: (16)_____ Average Requirements, (17)_____ Dietary Allowances, (18)_____ Intakes, and (19)_____ Upper Intake Levels.

Given too much, too little, or an imbalance of nutrients, the body becomes (20)_____. The sequence of events begins with poor nutrient (21)_____, or inefficient absorption or use of the nutrient, then progresses to internal abnormalities, and finally manifests itself in externally observable (22)_____.

ASSIGNMENTS

Answer these chapter study questions from the text:

1. Give several reasons (and examples) why people make the food choices that they do.

2. What is a nutrient? Name the six classes of nutrients found in foods. What is an essential nutrient?

3. Which nutrients are inorganic and which are organic? Discuss the significance of that distinction.

4. Which nutrients yield energy and how much energy do they yield per gram? How is energy measured?

5. Describe how alcohol resembles nutrients. Why is alcohol not considered a nutrient?

6. What is the science of nutrition? Describe the types of research studies and methods used in acquiring nutrition information.

7. Explain how variables might be correlational but not causal.

8. What are the DRI? Who develops the DRI? To whom do they apply? How are they used? In your description, identify the four categories of DRI and indicate how they are related.

9. What judgment factors are involved in setting the energy and nutrient intake recommendations?

10. What happens when people either get too little or too much energy or nutrients? Define malnutrition, undernutrition, and overnutrition. Describe the four methods used to detect energy and nutrient deficiencies and excesses.

11. What methods are used in nutrition surveys? What kinds of information can these surveys provide?

12. Describe risk factors and their relationships to disease.

CHAPTER GLOSSARY

Acceptable Macronutrient Distribution Ranges (AMDR): ranges of intakes for the energy nutrients that provide adequate energy and nutrients and reduce the risk of chronic diseases.
acute disease: a disease that develops quickly, produces sharp symptoms, and runs a short course.
Adequate Intake (AI): the average amount of a nutrient that appears sufficient to maintain a specified criterion.
anecdote: a personal account of an experience or event; not accepted as reliable scientific information.
anthropometric: relating to measurement of the physical characteristics of the body, such as height and weight.
atom: the smallest component of an element that has all of the properties of the element.
blind experiment: an experiment in which the subjects do not know whether they are members of the experimental or the control group.
calories: units by which energy is measured. Food energy is measured in kilocalories (1000 calories equal 1 kilocalorie), abbreviated kcalories or kcal. One kcalorie is the amount of heat necessary to raise the temperature of 1 kilogram (kg) of water 1° C.
chronic diseases: degenerative diseases characterized by deterioration of the body organs; diseases of long duration that progress slowly; also called *noncommunicable diseases* (NCD).
compound: a substance composed of two or more different atoms—for example, H_2O.
control group: a group of individuals similar in all possible respects to the group being experimented on except for the experimental treatment.
correlation: the simultaneous increase, decrease, or change of two variables.
covert: hidden, as if under covers.
deficient: the amount of a nutrient below which almost all healthy people can be expected, over time, to experience deficiency symptoms.
diet: the foods and beverages a person eats and drinks.
Dietary Reference Intakes (DRI): a set of values for the dietary nutrient intakes of healthy people in the United States and Canada; these include: Estimated Average Requirements, Recommended Dietary Allowances, Adequate Intakes, and Tolerable Upper Limits.
double-blind experiment: an experiment in which neither the subjects nor those conducting the experiment know which subjects are members of the experimental group and which are serving as control subjects, until after the experiment is over.
element: a substance composed of atoms that are alike--for example, iron (Fe).
energy: the capacity to do work. The energy in food is chemical energy. The body can convert this chemical energy to mechanical, electrical, or heat energy.
energy density: a measure of the energy a food provides relative to the amount of food (kcalories per gram).

energy-yielding nutrients: the nutrients that break down to yield energy the body can use: carbohydrate, fat, protein.
essential nutrients: nutrients a person must obtain from food because the body cannot make them for itself in sufficient quantity to meet physiological needs; also called *indispensable nutrients*.
Estimated Average Requirement: the amount of a nutrient that will maintain a specific biochemical or physiological function in half the people of a given age and sex group.
Estimated Energy Requirement (EER): the average dietary energy intake that maintains energy balance and good health in a person of a given age, gender, weight, height, and level of physical activity.
experimental group: a group of individuals similar in all possible respects to the control group except for the treatment. The experimental group receives the real treatment.
FAO: the Food and Agriculture Organization (of the United Nations).
food consumption survey: a survey that measures the amounts and kinds of foods people consume (using diet histories), estimates the nutrient intakes, and compares them with a standard such as the RDA.
foods: products derived from plants or animals that can be taken into the body to yield nutrients for the maintenance of life and the growth and repair of tissues.
functional foods: foods or food ingredients that have been modified to provide a health benefit beyond their nutrient contribution.
genome: the full complement of genetic material (DNA) in the chromosomes of a cell.
Healthy People: a national public health initiative that identifies the most significant threats to health and focuses on efforts toward eliminating them.
inorganic: not containing carbon or pertaining to living things.
malnutrition: any condition caused by excess or deficient food energy or nutrient intake or by an imbalance of nutrients.
minerals: inorganic elements; some minerals are essential nutrients required in small amounts.
molecule: two or more atoms of the same or different elements joined by chemical bonds.
nonnutrients: compounds in foods with no known nutritional value.
nutrients: substances obtained from food and used in the body to provide energy and structural materials and to regulate growth, maintenance, and repair of the body's tissues; nutrients may also reduce the risks of some chronic diseases.
nutrition: the science of foods and the nutrients and other substances they contain, and of their actions within the body.
nutrition assessment: a comprehensive approach, completed by a registered dietitian, to defining nutrition status that uses health, socioeconomic, drug, and diet histories; anthropometric measurements; physical examinations; and laboratory tests.
nutrition genomics: the science of how nutrients affect the activities of genes and how genes affect the activity of nutrients.
nutrition status survey: a survey that evaluates people's nutrition status using diet histories, anthropometric measures, physical examinations, and laboratory tests.
organic: a substance or a molecule containing carbon-carbon bonds or carbon-hydrogen bonds.
overnutrition: excess energy or nutrients.
overt: out in the open and easy to observe.
peer review: a process in which a panel of scientists rigorously evaluates a research study to assure that the scientific model was followed.
phytochemicals: nonnutritive compounds found in plant-derived foods that have biological activity in the body.
placebo: an inert, harmless medication given to provide comfort and hope.
placebo effect: the healing effect that faith in medicine, even inert medicine, often has.
primary deficiency: a nutrient deficiency caused by inadequate dietary intake of a nutrient.

randomization: a process of choosing the members of the experimental and control groups without bias.
Recommended Dietary Allowance (RDA): the average daily amount of a nutrient considered adequate to meet the known nutrient needs of practically all healthy people.
registered dietitian: a college-educated food and nutrition specialist who is qualified to evaluate people's nutritional health and needs.
replication: repeating an experiment and getting the same results.
requirement: the amount of a nutrient that will maintain normal biochemical and physiological functions and prevent the development of specific deficiency signs.
risk factors: factors associated with an elevated frequency of a disease but not proven to be causal.
science of nutrition: the study of the nutrients in foods and of the body's handling of them.
secondary deficiency: a nutrient deficiency caused by something other than diet, such as a disease condition that reduces absorption, accelerates use, hastens excretion, or destroys the nutrient.
subclinical deficiency: a deficiency in the early stages, before the outward signs have appeared.
subjects: people or animals participating in a research project.
Tolerable Upper Intake Level (UL): the maximum amount of a nutrient that appears safe for most healthy people and beyond which there is an increased risk of adverse health effects.
undernutrition: deficiency of energy or nutrients.
upper safe: the amount of a nutrient that appears safe for *most healthy people* and beyond which there is concern that some people will experience toxicity symptoms.
validity: having the quality of being founded on fact or evidence.
variable: a factor that changes.
vitamins: organic, essential nutrients required in small amounts by the body for health.
WHO: the World Health Organization.

Complete these short answer questions:

1. The six classes of nutrients are:

   a.   d.

   b.   e.

   c.   f.

2. Which nutrients are organic?

   a.   c.

   b.   d.

3. Which nutrients are inorganic?

   a.   b.

4. Which nutrients provide energy?

   a.   c.

   b.

5. 1 gram carbohydrate = _____ kcalories
    1 gram fat = _____ kcalories
    1 gram protein = _____ kcalories
    1 gram alcohol = _____ kcalories

6. 1/2 cup vegetables = _____ ml or _____ grams
    1/2 cup juice or milk = _____ ml or _____ grams
    1 teaspoon dry powder = _____ grams

7. The water-soluble vitamins are:

    a.   b.

8. The fat-soluble vitamins are:

    a.   c.
    b.   d.

9. Name and describe the methods used in nutrition assessment.

    a.
    b.
    c.
    d.

Solve these problems:

1. How many grams of fat were consumed, if a person received 405 kcalories from fat in a day?

2. How many kcalories are in 13 grams of carbohydrate?

3. How many total kcalories are in 10 grams of carbohydrate, 4 grams of protein, 6 grams of fat, and 7 grams of alcohol?

Complete this crossword puzzle by Mary A. Wyandt, Ph.D., CHES.

	Across		Down
1.	an experiment in which the subjects do not know whether they are members of the experimental or the control group	2.	the healing effect that faith in medicine, even inert medicine, often has
5.	repeating an experiment and getting the same results	3.	a group of individuals similar in all possible respects to the group being experimented on except for the experimental treatment
6.	the simultaneous increase, decrease, or change of two variables	4.	a factor that changes
7.	a group of individuals similar in all possible respects to the control group except for the treatment	8.	having the quality of being founded on fact or evidence
10.	a process of choosing the members of the experimental and control groups without bias	9.	people or animals participating in a research project

Sample Test Questions

1. When people eat foods of their families and geographical location, their eating patterns are being influenced by:

    a. personal preference.
    b. habit.
    c. ethnic heritage or tradition.
    d. physical appearance.
    e. nutrition.

2. People who eat foods to relieve boredom or calm anxiety are eating for reasons of:

    a. availability.
    b. nutrition.
    c. values.
    d. emotional comfort.
    e. economy.

3. The science of nutrition is the study of:

    a. how to prepare delicious foods that will improve a person's health.
    b. nutrient ingestion, digestion, absorption, and transport.
    c. nutrient metabolism, interaction, storage, and excretion.
    d. a and b.
    e. b and c.

4. A complete chemical analysis of your body would show that it is composed mostly of:

    a. proteins.
    b. carbohydrates.
    c. fats.
    d. water.
    e. minerals.

5. An organic compound is:

    a. a compound that contains carbon and hydrogen atoms.
    b. any substance found in living organisms.
    c. a compound that contains oxygen atoms.
    d. found only in foods grown under special conditions.
    e. superior.

6. Among these classes of nutrients, which one is not organic?

    a. carbohydrate
    b. fat
    c. protein
    d. vitamins
    e. minerals

7. Which does not yield energy for human use?

    a. carbohydrates
    b. vitamins
    c. fats
    d. proteins

8. When energy nutrients are metabolized:

    a. the arrangement of atoms remains unaltered.
    b. energy is released.
    c. the bonds between nutrient's atoms break.
    d. a and b.
    e. b and c.

9. A kcalorie is a:

    a. gram of fat.
    b. unit in which energy is measured.
    c. heating device.
    d. term used to describe the amount of sugar and fat in foods.

10. Any food can be "fattening" if you eat too much of it, even protein-rich food.

    a. True   b. False

11. A carbohydrate-rich food like bread:

    a. contains a mixture of the three energy nutrients.
    b. contains no protein.
    c. contains no protein or fat.
    d. cannot be properly classified on the basis of nutrient content.

12. One gram of alcohol contains:

    a. 4 kcals.   c. 9 kcals.
    b. 7 kcals.   d. 0 kcals.

13. Which of the following are fat-soluble?

    1. vitamin A   4. vitamin D
    2. B vitamins   5. vitamin E
    3. vitamin C   6. vitamin K

    a. 1, 4, 5, 6   d. 2, 3
    b. 2, 3, 5   e. 2, 5, 6
    c. 1, 4

14. Minerals are:

    a. energy providers.   c. large.
    b. inorganic elements.   d. organic elements.

15. Water:

    a. is organic.
    b. gives us energy.
    c. is indispensable.
    d. is not a nutrient.

16. This type of research study observes how much and what kinds of foods a group of people eat:

    a. human intervention
    b. animal
    c. case-control
    d. epidemiological

17. This is a set of four nutrient intake values that can be used to plan and evaluate diets for healthy people.

    a. RDA
    b. AI
    c. TUIL
    d. DRI

18. The dietary recommendations apply to:

    a. average daily intakes.
    b. healthy people.
    c. all people.
    d. a and b.
    e. a and c.

19. Historical information, physical examination, laboratory tests, anthropometric measures are:

    a. steps in the scientific method.
    b. methods used in nutrition assessment.
    c. procedures used to determine nutrient content of a diet.
    d. obsolete procedures used by nutritionists years ago.

20. A deficiency caused by an inadequate intake of a nutrient is a _____ deficiency.

    a. primary
    b. secondary
    c. dire
    d. clinical
    e. subclinical

21. A technique to detect nutrient deficiencies by taking height and weight measurements is part of the nutrition assessment component known as:

    a. diet history.
    b. anthropometrics.
    c. physical examination.
    d. biochemical tests.

22. Factors associated with elevated frequency of a disease but not proved to be causal are:

    a. medical factors
    b. individual interventions
    c. risk factors
    d. disease clusters

ANSWERS

Summing Up—(1) nutrition; (2) social interactions; (3) emotional comfort; (4) nutrients; (5) fat; (6) vitamins; (7) water; (8) organic; (9) energy; (10) nutrient; (11) growth; (12) essential; (13) kcalories; (14) work; (15) Dietary; (16) Estimated; (17) Recommended; (18) Adequate; (19) Tolerable; (20) malnourished; (21) intake; (22) symptoms.

Chapter study questions from the text—(1) Personal preference, habit, ethnic heritage or tradition, social interactions, availability, convenience, economy, positive and negative associations, emotional comfort, values, image, nutrition. (2) A substance obtained from food and used in the body to promote growth, maintenance, and repair. Carbohydrate, fat, protein, vitamins, minerals, and water. Essential: must be obtained from an outside source. (3) Organic nutrients: carbohydrates, fats, proteins, vitamins. Inorganic nutrients: minerals, water. Organic means all carbon compounds, not necessarily living things. (4) Energy yielding nutrients: carbohydrate (4 kcal/gm), fat (9 kcal/gm), protein (4 kcal/gm). Measured in calories or kilocalories. (5) Alcohol yields energy (7 kcals per gram) when metabolized but alcohol is not considered a nutrient because it does not support the growth, maintenance, or repair of the body. (6) The science of nutrition is the study of the nutrients in foods and the body's handling of those nutrients. Epidemiological studies, case-control studies, animal studies, human intervention (or clinical) trials. (7) Correlational variables are associated with each other; causal variables require a mechanism and indicate one causes the other. (8) The DRI are the Dietary Reference Intakes. They are a set of four nutrient intake values that can be used to plan and evaluate diets for healthy people. They are developed by the DRI Committee. Members of the committee are selected from the Food and Nutrition Board of the Institute of Medicine, the National Academy of Sciences, and Health Canada. The four categories include the Estimated Average Requirement (defines the amount of a nutrient that supports a specific function in the body for half of the population); the Recommended Dietary Allowance (uses Estimated Average Requirement to establish a goal for dietary intake that will meet the needs of almost all healthy people); an Adequate Intake (serve a similar purpose when an RDA cannot be determined); and Tolerable Upper Intake Level (establishes the highest amount that appears safe for regular consumption). (9) How much of a nutrient a person needs is determined by studying deficiency states, nutrient stores, and depletion, and by measuring the body's intake and excretion of the nutrient; that different individuals have different requirements; at what dividing line the bulk of the population is covered. (10) They get sick and show signs of deficiencies. Malnutrition is poor nutrition status, undernutrition is underconsumption of food energy or nutrients severe enough to cause disease or increased susceptibility to disease, and overnutrition is overconsumption of food energy or nutrients severe enough to cause disease or to cause increased susceptibility to disease. Through nutrition assessment techniques (anthropometric measures, lab tests, physical findings, and diet history). (11) Food consumption surveys and nutrition status surveys can provide information regarding amounts and kinds of foods people consume as well as evaluate people's nutrition status. (12) Factors that increase the risk of developing chronic diseases are called risk factors. A strong association between a risk factor and a disease means that when the factor is present, the likelihood of developing the disease increases.

Short Answers—
1. carbohydrate; fat; protein; vitamins; minerals; water
2. carbohydrate; fat; protein; vitamins
3. minerals; water
4. carbohydrate; fat; protein
5. 4; 9; 4; 7
6. 125; 100; 125; 100; 5

7. B vitamins; vitamin C
8. vitamin A; vitamin D; vitamin E; vitamin K
9. data on diet--record foods eaten over a period of time; physical examination--inspect body parts (hair, eyes, skin); laboratory tests--analyze body samples (blood, urine); anthropometric measures--measure height, weight, body parts

Problem Solving—
1. 45 g (405 kcal divided by 9 kcal/g fat = 45 g fat)
2. 52 kcal (13 g carb multiplied by 4 kcal/g carb = 52 kcal)
3. 159 kcal (10 g carb X 4 kcal/g + 4 g pro X 4 kcal/g + 6 g fat X 9 kcal/g + 7 g alc X 7 kcal/g = 40 + 16 + 54 + 49 = 159 kcal)

Crossword Puzzle—

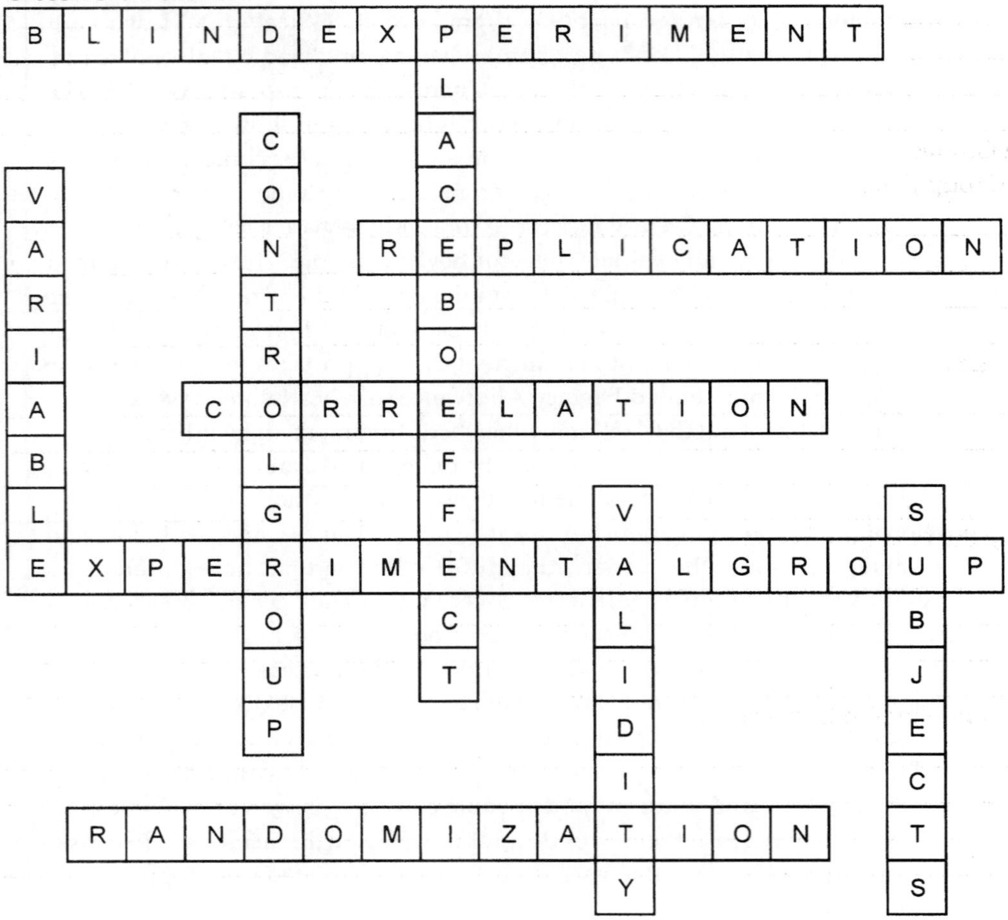

Across:
- BLINDEXPERIMENT
- REPLICATION
- CORRELATION
- EXPERIMENTALGROUP
- RANDOMIZATION

Down:
- VARIABLE
- CONTROLGROUP
- PLACEBO
- LACEBO (intersecting)
- SUBJECTS
- VALIDITY

Sample Test Questions—

1.	c (p. 2)	7.	b (p. 6-7)	13.	a (p. 7)	19.	b (p. 16-17)
2.	d (p. 3)	8.	e (p. 7)	14.	b (p. 8)	20.	a (p. 17)
3.	e (p. 9)	9.	b (p. 5)	15.	c (p. 8)	21.	b (p. 16)
4.	d (p. 5)	10.	a (p. 6-7)	16.	d (p. 9)	22.	c (p. 19)
5.	a (p. 5)	11.	a (p. 7)	17.	d (p. 12)		
6.	e (p. 5)	12.	b (p. 7)	18.	d (p. 12)		

Chapter 2
Planning a Healthy Diet

Chapter Outline

I. Principles and Guidelines
    A. Diet-Planning Principles

    B. Dietary Guidelines for Americans

II. Diet-Planning Guides
    A. Food Group Plans

    B. Exchange Lists

    C. Putting the Plan into Action

    D. From Guidelines to Groceries

III. Food Labels
    A. The Ingredient List

    B. Serving Sizes

C. Nutrition Facts

D. The Daily Values

E. Nutrient Claims

F. Health Claims

G. Structure-Function Claims

H. Consumer Education

Highlight: A World Tour of Pyramids, Pagodas, and Plates

Summing Up

Six concepts to keep in mind when planning a nutritious diet are: (1) _____, (2) _____, (3) _____ control, nutrient density, moderation, and (4) _____. The Dietary Guidelines encourage people to eat a (5) _____ of foods to get the (6) _____ needed to support good health and the energy appropriate to maintain a healthy (7) _____. Food group plans and (8) _____ systems use these concepts.

Food Group Plans divide foods into groups according to similarity of protein, (9)_____, and mineral content. A specific number of daily food (10)_____ is recommended from each group. The Daily Food Guide differs from the 4 food group plan by splitting (11)_____ from fruits.

Exchange systems list foods according to the number of kcalories and the amount of (12)_____, (13)_____, and (14)_____ each contains. Each list defines specific portion (15)_____ for individual foods. Exchange systems are useful for monitoring (16)_____ intakes.

When grocery shopping, it is wise to choose (17)_____-_____ breads and cereals, fresh (18)_____ and fruits and moderate amounts of lean meats and (19)_____ products. Current food labels provide consumers with useful information about the foods they eat, and how individual foods fit into their daily (20)_____.

ASSIGNMENTS

Answer these chapter study questions from the text:

1. Name the diet-planning principles and briefly describe how each principle helps in diet planning.

2. What recommendations appear in the *Dietary Guidelines for Americans*?

3. Name the five food groups in the Daily Food Guide and identify several foods typical of each group. Explain how such plans group foods and what diet-planning principles the plans best accommodate. How are food group plans used, and what are some of their strengths and weaknesses?

4. Review the *Dietary Guidelines*. What types of grocery selections would you make to achieve those recommendations?

5. What information can you expect to find on a food label? How can this information help you choose between two similar products?

6. What are the Daily Values? How can they help you meet health recommendations?

7. What health claims have been approved by FDA for use on labels? What criteria must all health claims meet?

CHAPTER GLOSSARY

adequacy (dietary): providing all the essential nutrients, fiber, and energy in amounts sufficient to maintain health.
balance (dietary): providing foods of a number of types in proportion to each other, such that foods rich in some nutrients do not crowd out of the diet foods that are rich in other nutrients.
bran: the protective coating around the kernel similar in function to the shell of a nut, rich in nutrients and fiber.
Daily Reference Values (DRV): a set of standards for nutrients and food components (such as fat and fiber) that have important relationships with health; used on food labels as part of the Daily Values.
Daily Values (DV): reference values developed by the FDA specifically for use on food labels. The Daily Values represent two sets of standards: Reference Daily Intakes (RDI) and Daily Reference Values (DRV).
empty-kcalorie foods: a popular term used to denote foods that contribute energy but lack protein, vitamins and minerals.
endosperm: the bulk of the edible part of the kernel containing starch and proteins.
enriched: addition of nutrients to a food; adding nutrients that were lost during processing so that the food will meet a specified standard.
exchange lists: diet-planning tools that organize foods by their proportions of carbohydrate, fat, and protein. Foods on any single list can be used interchangeably.
food group plans: diet-planning tools that sort foods of similar origin and nutrient content into groups and then specify that people should eat certain numbers of servings from each group.
food substitutes: foods that are designed to replace other foods.
fortified: the addition of nutrients that were either not originally present or present in insignificant amounts to a food.
germ: the nutrient-rich inner part of a grain. The germ is the seed that grows into a wheat plant, so it is especially rich in vitamins and minerals to support new life.
gluten: an elastic protein found in wheat and other grains that gives dough its structure and cohesiveness.
health claims: statements that characterize the relationship between any nutrient or other substance in a food and a disease or health-related condition.

Healthy Eating Index: a measure developed by the USDA for assessing how well a diet conforms to the recommendations of the Food Guide Pyramid and the Dietary Guidelines for Americans.

husk: the outer, inedible part of a grain; also called the *chaff*.

imitation food: a food that substitutes for and resembles another food and is nutritionally inferior to it with respect to vitamin, mineral, or protein content.

kcalorie (energy) control: management of food energy intake.

legumes: plants of the bean and pea family, rich in high-quality protein compared with other plant-derived foods.

moderation (dietary): in relation to dietary intake, providing enough but not too much of a substance.

nutrient claims: statements that characterize the quantity of a nutrient in a food.

nutrient density: a measure of the nutrients a food provides relative to the energy it provides. The more nutrients and the fewer kcalories, the higher the nutrient density.

Reference Daily Intakes (RDI): a set of standards for protein, vitamins and minerals used on food labels as part of the Daily Values; previously known as the U.S. RDA.

refined: the process by which the coarse parts of a food are removed. With respect to refining wheat into flour, the bran, germ, and husk have been removed, leaving only the endosperm.

structure-function claims: statements that characterize the relationship between a nutrient or other substance in a food and its role in the body.

substitute food: a food that is designed to replace another.

textured vegetable protein: processed soybean protein used in vegetarian products such as soy burgers.

unbleached flour: a tan-colored endosperm flour with texture and nutritive qualities that approximate those of regular white flour.

variety (dietary): eating a wide selection of foods within and among the major food groups.

wheat flour: any flour made from wheat, including white flour.

white flour: an endosperm flour that has been refined and bleached for maximum softness and whiteness.

whole grain: a grain milled in its entirety (all but the husk), not refined.

whole-wheat flour: flour made from whole-wheat kernels; a whole-grain flour.

Complete these short answer questions:

1. Diet-planning principles include:

    a.

    b.

    c.

    d.

    e.

    f.

2. Two diet-planning guides most widely used are:

    a.

    b.

3. The four groups in the Four Food Group Plan are:

   a.   c.

   b.   d.

4. The five groups in the Daily Food Guide are:

   a.

   b.

   c.

   d.

   e.

5. The six lists of foods in the exchange system and their kcalorie values per single portion sizes are:

   a.   d.

   b.   e.

   c.   f.

Solve this problem:

1. Using the exchange system, how many kcalories are in the following breakfast?

   1 slice of whole-wheat toast with
   1 tsp butter
   1 small banana, ½ cup apple juice
   1 ½ cups nonfat milk

Complete this crossword puzzle by Mary A. Wyandt, Ph.D., CHES.

Across	Down
1. statements that characterize the quantity of a nutrient in a food.	2 a tan-colored endosperm flour with texture and nutritive qualities that approximate those of regular white flour
3. an elastic protein found in wheat and other grains that gives dough its structure and cohesiveness	4 plants of the bean and pea family, rich in high-quality protein compared with other plant-derived foods
7. the outer, inedible part of a grain; also called the chaff	5 a grain milled in its entirety (all but the husk); not refined
8. the bulk of the edible part of the kernel containing starch and protein	6 the protective coating around the kernel similar in function to the shell of a nut, rich in nutrients and fiber
10. a measure of the nutrients a food provides relative to the energy it provides	9 the nutrient-rich inner part of a grain; the seed that grows into a wheat plant

SAMPLE TEST QUESTIONS

1. Diet planning principles include:

    a. adequacy, B vitamins, carbohydrates, moderation, nutrient density, variety.
    b. abundance, balance, carbohydrates, meals, nutrients, vegetables.
    c. adequacy, balance, kcalorie control, nutrient density, moderation, variety.
    d. abundance, B vitamins, kcalorie control, milk, moderation, vegetables.

2. Selecting foods that deliver the most nutrients for the least food energy is applying the concept of:

    a. moderation
    b. abundance
    c. variety
    d. adequacy
    e. nutrient density

3. Which of the following is the most nutrient dense food relative to calcium content?

    a. whole milk
    b. non-fat milk
    c. low-fat milk
    d. cheddar cheese

4. An adult following the Food Guide Pyramid should consume at least _____ serving(s) of vegetables each day.

    a. one
    b. two
    c. three
    d. four
    e. five

5. Which of the following is descriptive of the USDA Food Pyramid?

    a. a three-dimensional structure designed to assist the average consumer in the use of the Food Exchange System.
    b. an education tool for teaching nutrition to children that consists of food blocks that require stacking in a specific order.
    c. a graphic representation of the Dietary Guidelines that displays complex carbohydrates at the base and fats and sweets at the very top.
    d. a system of specialized containers of several different sizes which allows for better storage and preservation of perishable food items.

6. The Daily Food Guide suggests _____ servings per day of bread and cereals.

    a. 6 to 11
    b. 3 to 5
    c. 2 to 4
    d. 2 to 3

7. Exchange lists group foods according to:

   1. protein, fat, and carbohydrate content.
   2. kcaloric content.
   3. vitamin content.
   4. mineral content.

   a. 1 only
   b. 2 only
   c. 3 and 4
   d. 1, 2, 3, 4

8. A measure used to assess how well a diet conforms to the recommendations of the Food Guide Pyramid and the Dietary Guidelines for Americans is called:

   a. Exchange Lists.
   b. Healthy People 2010 Objectives.
   c. Healthy Eating Index.
   d. Sample Diet Plan.

9. Which of the following is a disadvantage of the Food Guide Pyramid?

   a. It is used primarily by diabetics.
   b. It defines food portions according to energy content.
   c. The groupings are made without distinguishing among the individual foods within each group.
   d. They subdivide the grain group based on fiber content.

10. Which of these diets receives the most points on the Healthy Eating Index?

    a. 6 grains
    b. 1 vegetable
    c. 1 fruit
    d. 1 ounce of lean meat

11. Which of the following is a characteristic of enriched grain products?

    a. They have all of the added nutrients listed on the label.
    b. They have the fiber restored from the refining procedure.
    c. They have 4 vitamins and 4 minerals added by the food processor.
    d. They have virtually all the nutrients restored from the refining procedure.

12. When shopping for grains, this type of product may be rich in all nutrients found in original grain:

    a. refined
    b. enriched
    c. whole grain
    d. fiber rich

13. Dried beans and peas, pinto beans, lima beans and black beans are examples of:

    a. legumes
    b. meats
    c. milk alternatives
    d. fruits

14. Whole-grain bread contains more of the following nutrients than enriched bread:

    a. iron, carbohydrate, protein
    b. iron, thiamin, riboflavin, niacin
    c. magnesium, zinc, vitamin B$_6$, folate
    d. water, fiber, fat

15. The National Cancer Institute Recommends how many servings of vegetables each day?

    a. 1-3
    b   3-4
    c. 5-9
    d. 10-12
    e. 13 or more

16. Foods that substitute for and resemble another food, but are nutritionally inferior to it are called:

    a. functional.
    b. food substitutes.
    c. food density.
    d. imitation foods.

17. The items on a food label that tell consumers about the nutritional value of a product include:

    1. the common or usual name of the product
    2. the net contents in terms of weight measure or count
    3. the ingredients in descending order of predominance
    4. the serving size and number of servings per container
    5. the quantities of specified nutrients and food constituents

    a. 1, 2, 3, 4.
    b. 2, 3, 4, 5.
    c. 3, 4, 5.
    d. all of the above

18. Serving sizes on a label:

    a. vary according to brand.
    b. are not always the same as those of the Food Guide Pyramid.
    c. are 1 cup for all ice creams.
    d. b and c

19. The FDA uses _____ kcalories as a standard for energy intake in calculating the Daily Values (DV) for energy-yielding nutrients.

    a. 1,200 kcals
    b. 1,500 kcals
    c. 1,800 kcals
    d. 2,000 kcals

20. Health claims on a label are defined as:

   a. statements that characterize the relationship between a nutrient or other substance in a food and a disease or health-related condition.
   b. statements where scientifically valid links between diet and health have been clearly established.
   c. statements that emphasize the importance of the total diet in disease prevention.
   d. none of the above

ANSWERS

Summing Up—(1) adequacy; (2) balance; (3) kcalorie; (4) variety; (5) variety; (6) nutrients; (7) weight; (8) exchange; (9) vitamin; (10) servings; (11) vegetables (12) protein; (13) carbohydrate; (14) fat; (15) sizes; (16) kcalorie; (17) whole-grain; (18) vegetables; (19) milk; (20) diets.

Chapter study questions from the text—(1) *Adequacy*—providing all essential nutrients. *Balance*—providing foods of a number of types in proportion to each other so that foods rich in some nutrients do not crowd out foods rich in other nutrients. *Kcalorie control*—management of food energy intake. *Nutrient density*—selecting foods with high nutrient value and food energy. *Moderation*—providing enough but not too much of a dietary constituent. *Variety*—using different foods on different occasions; variety helps ensure adequacy and balance. (2) See Figure 2-1. (3) See Figure 2-2 for food groups and foods typical of each group. Groups foods similar in origin and that make notable contributions of the same key nutrients. Best accommodates adequacy and variety. Strengths: helps avoid undernutrition; weaknesses: possible overnutrition by emphasizing minimum number of servings. (4) See Figure 2-1 and Figure 2-2 (answers are individual). (5) The common or usual name of the product; the name and address of the manufacturer, packer, or distributor; the net contents in terms of weight, measure, or count; the ingredients in descending order of predominance by weight; the serving size and number of servings per container; the quantities of specified nutrients and food constituents. Consumers can use ingredient lists to compare nutrient density of products. (6) Daily Values are reference values developed by the FDA for use on food labels; they help people compare foods to their recommended intakes. (7) Calcium and osteoporosis; sodium and hypertension; dietary saturated fat and cholesterol and risk of coronary heart disease; dietary fat and cancer; fiber-containing grain products, fruits, and vegetables, and grain products that contain fiber, particularly soluble fiber, and risk of coronary heart diseases; fruits and vegetables and cancer. Food making a health claim must be a naturally good source of at least one of following nutrients: vitamin A, vitamin C, iron, calcium, protein, or fiber; and foods cannot contain more than 20% of Daily Value for total fat, saturated fat, cholesterol or sodium.

Short Answers—
1. adequacy, balance, kcalorie control, nutrient density, moderation, variety
2. food group plans; exchange systems
3. meat/meat alternates; milk/milk substitutes; fruits and vegetables; grains
4. breads and cereals, vegetables, fruits, meats and meat alternatives, milk and milk products
5. starch/bread (80 kcal); meat/meat alternates (very lean=35 kcal, lean=55 kcal, medium-fat=75 kcal, high-fat=100 kcal); vegetable (25 kcal); fruit (60 kcal); milk (non-fat=90 kcal, low-fat=120 kcal, whole=150 kcal); fat (45 kcal)

Problem Solving—
1. 80 + 45 + 2(60) + 1.5(90) = 380 kcal

Crossword Puzzle—

Across: NUTRIENTCLAIMS, GLUTEN, HUSK, ENDOSPERM, NUTRIENTDENSITY

Down (letters visible): UNBLEACHEDFLOUR, BRAN, LEGUME, WHOLEGRAIN, GERM

Sample Test Questions--

1. c (p. 40, 41)
2. e (p. 40)
3. b (p. 40)
4. c (p. 43)
5. c (p. 46)
6. a (p. 43)
7. a (p. 48)
8. c (p. 48)
9. c (p. 48)
10. a (p. 49)
11. a (p. 52)
12. c (p. 52)
13. a (p. 46)
14. c (p. 53)
15. c (p. 54)
16. d (p. 55)
17. c (p. 56, 57)
18. b (p. 57)
19. d (p. 58)
20. a (p. 60)

Chapter 3
Digestion, Absorption, and Transport

Chapter Outline

I. Digestion
    A. Anatomy of the Digestive Tract

    B. The Muscular Action of Digestion

    C. The Secretions of Digestion

    D. The Final Stage

II. Absorption
    A. Anatomy of the Absorptive System

    B. A Closer Look at the Intestinal Cells

III. The Circulatory Systems
    A. The Vascular System

    B. The Lymphatic System

IV. Regulation of Digestion and Absorption
   A. Gastrointestinal Hormones and Nerve Pathways

_____
_____
_____

   B. The System at Its Best

_____
_____
_____

Highlight: Common Digestive Problems

_____
_____
_____
_____

SUMMING UP

The gastrointestinal tract is a flexible muscular tube measuring about (1)_____ feet in length. Substances that penetrate the tract's wall will enter the body; but many things pass through unabsorbed. Food enters the mouth past epiglottis to (2)_____ through cardiac sphincter to stomach through (3)_____ to small intestine (duodenum, with entrance from gallbladder and (4)_____; then jejunum; then ileum) through ileocecal valve to large intestine past (5)_____ to rectum ending at anus.

Involuntary muscles and (6)_____ of the digestive tract function without any conscious effort. Peristalsis and (7)_____ mix and move intestinal contents along the gastrointestinal tract. Digestive (8)_____ cause food substances to break down into simpler compounds. Nutrients are then (9)_____ in the microvilli of the (10)_____ intestine. The nutrients are transported to any part of the body by the circulatory systems—the (11)_____ system and the lymphatic system. The (12)_____ and nervous systems coordinate the digestive and absorptive processes. Characteristics of meals that promote optimal absorption of nutrients are balance, (13)_____, adequacy, and (14)_____.

Assignments

Answer these chapter study questions from the text:

1. Describe the obstacles associated with digesting food and the solutions offered by the human body.

2. Describe the path food follows as it travels through the digestive system. Summarize the muscular actions that take place along the way.

3. Name five organs that secrete digestive juices. How do the juices and enzymes facilitate digestion?

4. Describe the problems involved with absorbing nutrients and the solutions offered by the small intestine.

5. How is blood routed through the digestive system? Which nutrients enter the bloodstream directly? Which are first absorbed into the lymph?

6. Describe how the body coordinates and regulates the processes of digestion and absorption.

7. How does the composition of the diet influence the functioning of the GI tract?

8. What steps can you take to help your GI tract function at its best?

Chapter Glossary

absorption: passage of nutrients from the GI tract into either the blood or the lymph.
anus: the terminal sphincter of the GI tract.
appendix: a narrow blind sac extending from the beginning of the colon; a vestigial organ with no known function.
arteries: vessels that carry blood away from the heart.
-ase: a word ending denoting an enzyme. Enzymes are often identified by the place they come from and the compounds they work on.
bicarbonate: an alkaline secretion of the pancreas, part of the pancreatic juice.
bile: an emulsifier that prepares fats and oils for digestion; an exocrine secretion made by the liver, stored in the gallbladder, and released into the small intestine when needed.
bolus: a portion; with respect to food, amount swallowed at one time.
capillaries: small vessels that branch from an artery and connect arteries to veins.
carbohydrase: an enzyme that hydrolyzes carbohydrates.
cardiac sphincter: the sphincter muscle at the junction between the esophagus and the stomach.
catalyst: a compound that facilitates chemical reactions without itself being changed in the process.
cholecystokinin or CCK: a hormone produced by cells of the intestinal wall. Target organ: the gallbladder. Response: release of bile and slowing of GI motility.
chyme: the semiliquid mass of partly digested food expelled by the stomach into the duodenum.
colon: the lower portion of the intestine that completes the digestive process; its segments are the ascending colon, the transverse colon, the descending colon, and the sigmoid colon.
crypts: tubular glands that lie between the intestinal villi and secrete intestinal juices into the small intestine.
digestion: the process by which food is broken down into absorbable units.
digestive enzymes: proteins found in digestive juices that act on food substances, causing them to break down into simpler compounds.
digestive system: all the organs and glands associated with the ingestion and digestion of food.
duodenum: the top portion of the small intestine.
emulsifier: a substance with both water-soluble and fat-soluble portions that promotes the mixing of oils and fats in a watery solution.
endocrine glands: a cell or group of cells that secrete their materials into the blood.
enterogastrones: gastrointestinal hormones.
epiglottis: cartilage in the throat that guards the entrance to the trachea and prevents fluid or food from entering it when a person swallows.
esophagus: the food pipe; the conduit from the mouth to the stomach.
exocrine glands: glands that secrete their materials into the digestive tract or onto the surface of the skin.
feces: waste matter discharged from the colon; also called stools.
gallbladder: the organ that stores and concentrates bile. When it receives the signal that fat is present in the duodenum, the gallbladder contracts and squirts bile through the bile duct into the duodenum.
gastric glands: exocrine glands in the stomach wall that secrete gastric juice into the stomach.
gastric juice: the digestive secretion of the gastric glands of the stomach.
gastric-inhibitory peptide: a hormone produced by the intestine. Target organ: the stomach. Response: slowing of the secretion of gastric juices and of GI motility.
gastrin: a hormone secreted by cells in the stomach wall. Target organ: the stomach. Response: secretion of gastric juice.
GI tract: the gastrointestinal tract or digestive tract; the principal organs are the stomach and intestines.
gland: a cell or group of cells that secretes materials for special uses in the body.

goblet cells: cells of the GI tract and lungs that secrete mucus.
hepatic vein: the vein that collects blood from the liver capillaries and returns it to the heart.
homeostasis: the maintenance of constant internal conditions (such as blood chemistry, temperature, and blood pressure) by the body's control systems.
hormones: chemical messengers. Hormones are secreted by a variety of glands in response to altered conditions in the body.
hydrochloric acid: an acid composed of hydrogen and chloride atoms (HCl). The gastric glands normally produce this acid.
hydrolysis: a chemical reaction in which a major reactant is split into two products, with the addition of H to one and OH to the other (from water).
ileocecal valve: the sphincter separating the small and large intestines.
ileum: the last segment of the small intestine.
intestinal flora: the bacterial inhabitants of the GI tract.
jejunum: the first two-fifths of the small intestine beyond the duodenum.
large intestine: the lower portion of intestine that completes the digestive process; its segments are the ascending colon, the transverse colon, the descending colon, and the sigmoid colon; also called *colon*.
lipase: an enzyme that hydrolyzes lipids (fats).
liver: the organ that manufacturers bile and is the first to receive nutrients from the intestines.
lymph: a clear yellowish fluid that resembles blood without the red blood cells; lymph transports fat from the GI tract; transports fat and fat-soluble vitamins to the bloodstream via lymphatic vessels.
lymphatic system: a loosely organized system of vessels and ducts that convey fluids toward the heart.
microvilli: tiny, hairlike projections on each cell of every villus that can trap nutrient particles and transport them into the cells; singular *microvillus*.
motility: the ability of the GI tract muscles to move.
mucus: a slippery substance secreted by goblet cells of the GI lining (and other body linings) that protects the cells from exposure to digestive juices (and other destructive agents).
pancreas: a gland that secretes digestive enzymes and juices into the duodenum.
pancreatic juice: the exocrine secretion of the pancreas, containing enzymes for the digestion of carbohydrate, fat, and protein as well as bicarbonate, a neutralizing agent.
peristalsis: wavelike muscular contractions of the GI tract that push its contents along.
pH: the unit of measure expressing a substance's acidity or alkalinity.
portal vein: the vein that collects blood from the GI tract and conducts it to capillaries in the liver.
protease: an enzyme that hydrolyzes proteins.
pyloric sphincter: the circular muscle that separates the stomach from the small intestine and regulates the flow of partially digested food into the small intestine.
rectum: the muscular terminal part of the intestine, extending from the sigmoid colon to the anus.
reflux: a backward flow.
saliva: the secretion of the salivary glands; the principal enzyme that begins carbohydrate digestion.
salivary glands: exocrine glands that secret saliva into the mouth.
secretin: a hormone produced by cells in the duodenum wall. Target organ: the pancreas. Response: secretion of bicarbonate-rich pancreatic juice.
segmentation: a periodic squeezing or partitioning of the intestine at intervals along its length by its circular muscles.
small intestine: a 10-foot length of small-diameter intestine that is the major site of digestion of food and absorption of nutrients; its segments are the duodenum, jejunum, and ileum.
sphincter: a circular muscle surrounding, and able to close, a body opening.
stomach: a muscular, elastic, saclike portion of the digestive tract that grinds and churns swallowed food, mixing it with acid and enzymes to form chyme.

stools: waste matter discharged from the colon; also called *feces*.
subclavian vein: connects the thoracic duct with the right upper chamber of the heart, providing a passageway by which lymph can be returned to the vascular system.
thoracic duct: the duct that conveys lymph toward the heart.
trachea: the windpipe; the passageway from the mouth and nose to the lungs.
veins: vessels that carry blood back to the heart.
villi: fingerlike projections from the folds of the small intestine; singular *villus*.

Complete this short answer question:

1. Some problems involved in digestion include:

    a.

    b.

    c.

    d.

    e.

    f.

    g.

Complete this crossword puzzle by Mary A. Wyandt, Ph.D., CHES.

Across	Down
1. an enzyme that hydrolyzes proteins	2. passage of nutrients from the GI tract into either the blood or lymph
4. a substance with both water-soluble and fat-soluble portions that promotes the mixing of oils and fats in a watery solution	3. a chemical reaction in which a major reactant is split into two products, with the addition of H to one and OH to the other (from water)
6. a hormone secreted by cells in the stomach wall	5. an enzyme that hydrolyzes lipids (fats)
7. a portion; with respect to food, amount swallowed at one time	7. an emulsifier that prepares fats and oils for digestion; an exocrine secretion made by the liver and stored in the gallbladder
8. a hormone produced by cells in the duodenum wall	9. a backward flow

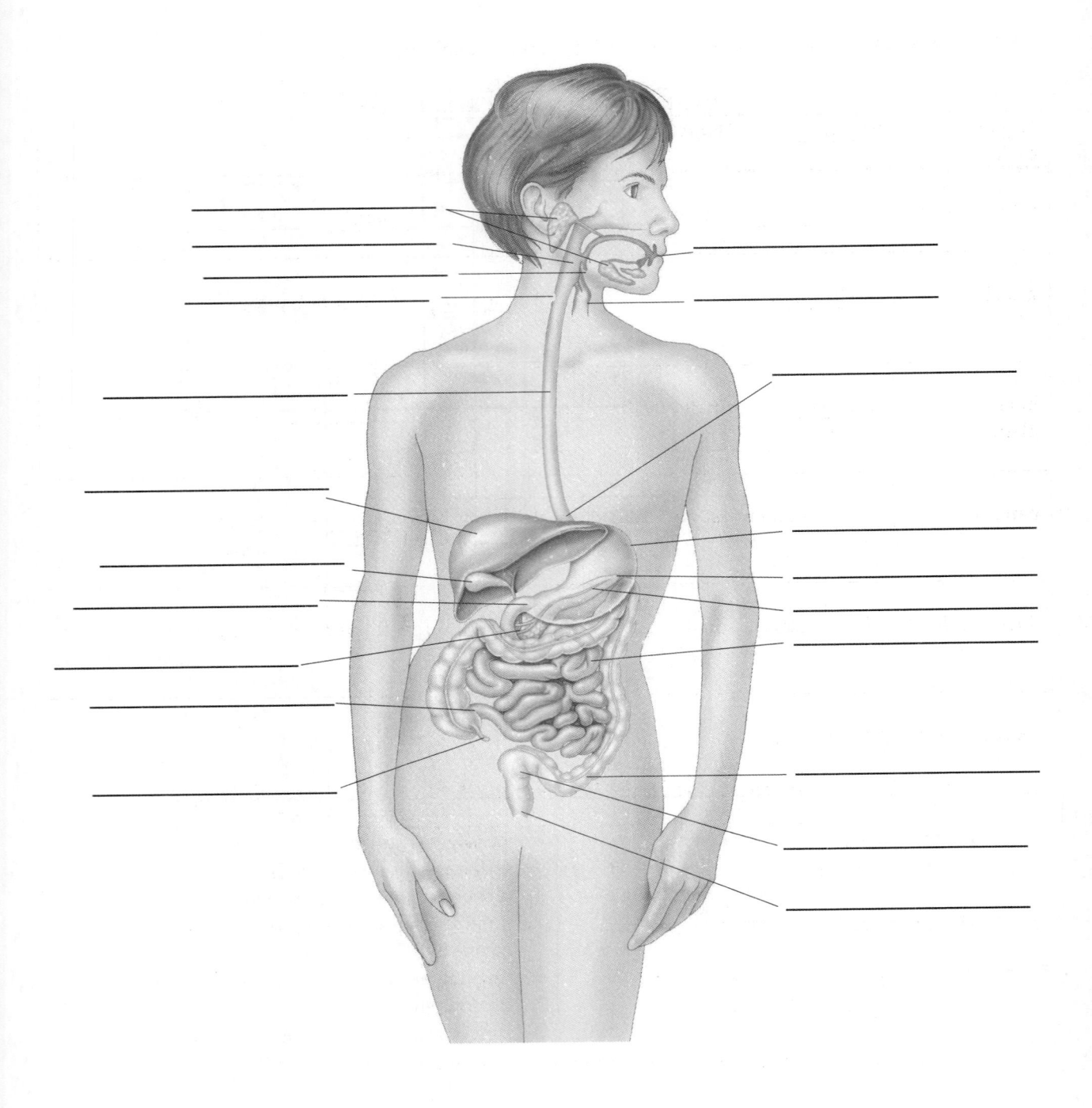

Solve these problems:
1. Identify these parts of the GI tract on the figure (above).

Anus	Large intestine (colon)	Pyloric sphincter
Appendix	Liver	Rectum
Bile duct	Lower esophageal sphincter	Small intestine
Epiglottis	Mouth	Salivary glands
Esophagus	Pancreas	Stomach
Gallbladder	Pancreatic duct	Trachea
Ileocecal valve	Pharynx	Upper esophageal sphincter

2. Identify the digestive secretions and their primary actions in this table.

Digestive Secretions and Their Actions

Organ or Gland	Secretion → Target	Action
salivary glands	? → mouth	starch —amylase→ ?
gastric glands	gastric juice → ?	? —pepsin/HCl→ smaller polypeptides
intestinal glands	intestinal juice → small intestine	carbohydrate —?→ monosaccharides
pancreas	? juice → small intestine	protein —protease→ dipeptides tripeptides ?
liver —bile→ ? —bile→ small intestine		fats —bile→ (emulsified fat) —?→ monoglycerides ? and fatty acids

SAMPLE TEST QUESTIONS

1. The muscular contractions that propel food through the digestive tract are called:

    a. defecation.
    b. peristalsis.
    c. hydrolysis.
    d. denaturation.

2. Which sphincter muscle is situated between the stomach and the small intestine?

    a. cardiac sphincter.
    b. pyloric sphincter.
    c. ileocecal valve.
    d. rectal sphincter.

3. The partially digested food that enters the small intestine from the stomach is called:

    a. micelle.
    b. bile.
    c. chyme.
    d. feces.

4. An enzyme that hydrolyzes proteins is called:

    a. bile.
    b. sphincter valve.
    c. pancreatic amylase.
    d. protease.
    e. secretory peptide.

5. The main function of bile is to:

   a. emulsify fats.
   b. stimulate the activity of protein digestive enzymes.
   c. neutralize the contents of the intestine.
   d. increase peristalsis.
   e. a, b, and c.

6. Nutrients that are digested in the small intestine are:

   a. carbohydrate, fat, and protein.
   b. fat, water, and fiber.
   c. protein, vitamins, and fiber.
   d. water, fiber, and minerals.

7. A narrow blind sac extending from the beginning of the colon is called:

   a. epiglottis   c. rectum
   b. anus   d. appendix

8. A single villus is composed of hundreds of _____ each covered with _____.

   a. organisms; mucus   d. cells; microvilli
   b. villi; lymph   e. crypts; glands
   c. enzymes; mucus

9. A group of cells that secretes materials for special uses in the body is:

   a. a gland   c. an enzyme
   b. a sphincter   d. bile

10. People should not eat certain food combinations at the same meal because the digestive system cannot handle more than one task at a time.

    a. true   b. false

11. A periodic squeezing of the intestines is:

    a. peristalsis   c. segmentation
    b. diverticulosis   d. defecation

12. The small intestine is about _____ feet long.

    a. 10
    b. 30
    c. 50
    d. 100

13. Saliva contains:

    a. protease, bicarbonate, and water.
    b. bile, HCl, and water.
    c. chyme, pepsin, and bicarbonate.
    d. salts, carbohydrases, and water.
    e. a and b

14. The lymphatic system:

    a. contains red blood cells.
    b. eventually drains into the blood circulatory system.
    c. collects in a large duct behind the heart.
    d. a and b
    e. b and c

15. These cells produce mucus that protects the stomach walls from gastric juices.

    a. secretin
    b. bile
    c. cholecystokinin
    d. goblet
    e. enterogastrone

16. This hormone is released in the presence of fat and slows intestinal motility to allow a longer digestion time.

    a. secretin
    b. bile
    c. cholecystokinin
    d. gastrin
    e. enterogastrone

17. The greatest danger from prolonged vomiting is:

    a. exhaustion.
    b. excess loss of fluid and electrolytes.
    c. vitamin deficiencies.
    d. starvation.

18. A sensible idea for preventing constipation is to:

    a. take medication on a regular basis.
    b. cut down on water intake.
    c. include more high-fiber foods in the diet.
    d. include fewer high-fiber foods in the diet.

Answers

Summing Up—(1) 26; (2) esophagus; (3) pylorus; (4) pancreas; (5) appendix; (6) glands; (7) segmentation; (8) enzymes; (9) absorbed; (10) small; (11) vascular; (12) endocrine; (13) variety; (14) moderation.

Chapter study questions from the text—(1) The epiglottis closes off the airways so that food and liquid do not enter the lungs. Food passes through the diaphragm to reach the stomach by way of the esophagus. Fluids are added to the food as it travels through the system allowing smooth passage. Water is reabsorbed in the large colon, thus conserving water and creating a semisolid waste. Peristalsis keeps the materials steadily moving through the system and sphincter muscles serve as one-way gates allowing small quantities to pass at appropriate intervals. Stomach cells secrete mucus to protect them from acid and enzymes that would digest them. Rectal muscles prevent elimination until voluntarily performed. (2) Food enters the mouth and travels past the epiglottis, down the esophagus and through the cardiac sphincter to the stomach, then through the pyloric sphincter to the small intestine, on through the ileocecal valve to the large intestine, past the appendix to the rectum, ending at the anus. Muscular actions include: chewing, swallowing, peristalsis, segmentation, and sphincter contractions. (3) Salivary glands, stomach glands, pancreas, liver, gallbladder, intestinal glands. Salivary glands secrete saliva that contains amylase enzyme that breaks down starch; gastric juice is secreted by the cells in the stomach wall and contains pepsin and HCl that breaks down proteins; pancreatic juice contains bicarbonate that neutralized acidic gastric juices as well as other enzymes that break down CHO, protein and fat; the gallbladder secretes bile which emulsifies fat. (4) The body must find a way to absorb many molecules. It solves this by its anatomy—it has hundreds of folds, each covered with thousands of villi, which in turn are composed of hundreds of cells, which in turn are covered with microvilli, providing a very large surface area for nutrient molecules to make contact and be absorbed. (5) Heart to arteries to capillaries (in intestines) to vein to capillaries (in liver) to vein to heart. Water-soluble nutrients and small products of fat digestion enter the bloodstream directly; large fats and fat-soluble nutrients are first absorbed into the lymph. (6) The body's hormonal system and nervous system coordinate all the digestive and absorptive processes. The contents in the GI tract either stimulate or inhibit digestive secretions by way of messages that are carried from one section of the GI tract to another by both hormones and nerve pathways. (7) Enzyme activity changes proportionately in response to the amounts of carbohydrate, fat and protein in the diet. Hormones in the GI tract inform the pancreas as to the amount and type of enzymes to secrete in response to diet; the presence of fat slows GI motility. (8) Obtain adequate sleep, physical activity, positive state of mind, meals with these characteristics: balance, moderation, variety, adequacy.

Short Answers—(1) a. Air and food must both go to the stomach but food and liquid cannot go to the lungs; b. Food must be conducted through the diaphragm to reach the abdomen; c. The amount of water should be regulated to keep the intestinal contents at the right consistency; d. Water must be withdrawn from the intestinal contents after absorption; e. The materials within the tract should move steadily except when a poison has been swallowed (in this case the contents should reverse direction and move quickly). If infection sets in farther down the tract, the flow should be accelerated; f. The cells of the digestive tract need protection against the powerful juices they secrete; g. Provision must be made for periodic, voluntary evacuation when convenient.

Crossword Puzzle—(1) protease; (2) absorption; (3) hydrolysis; (4) emulsifier; (5) lipase; (6) gastrin; (7D) bile; (7A) bolus; (8) secretin; (9) reflux

Problem Solving—
1. The Gastrointestinal Tract (see figure on next page)

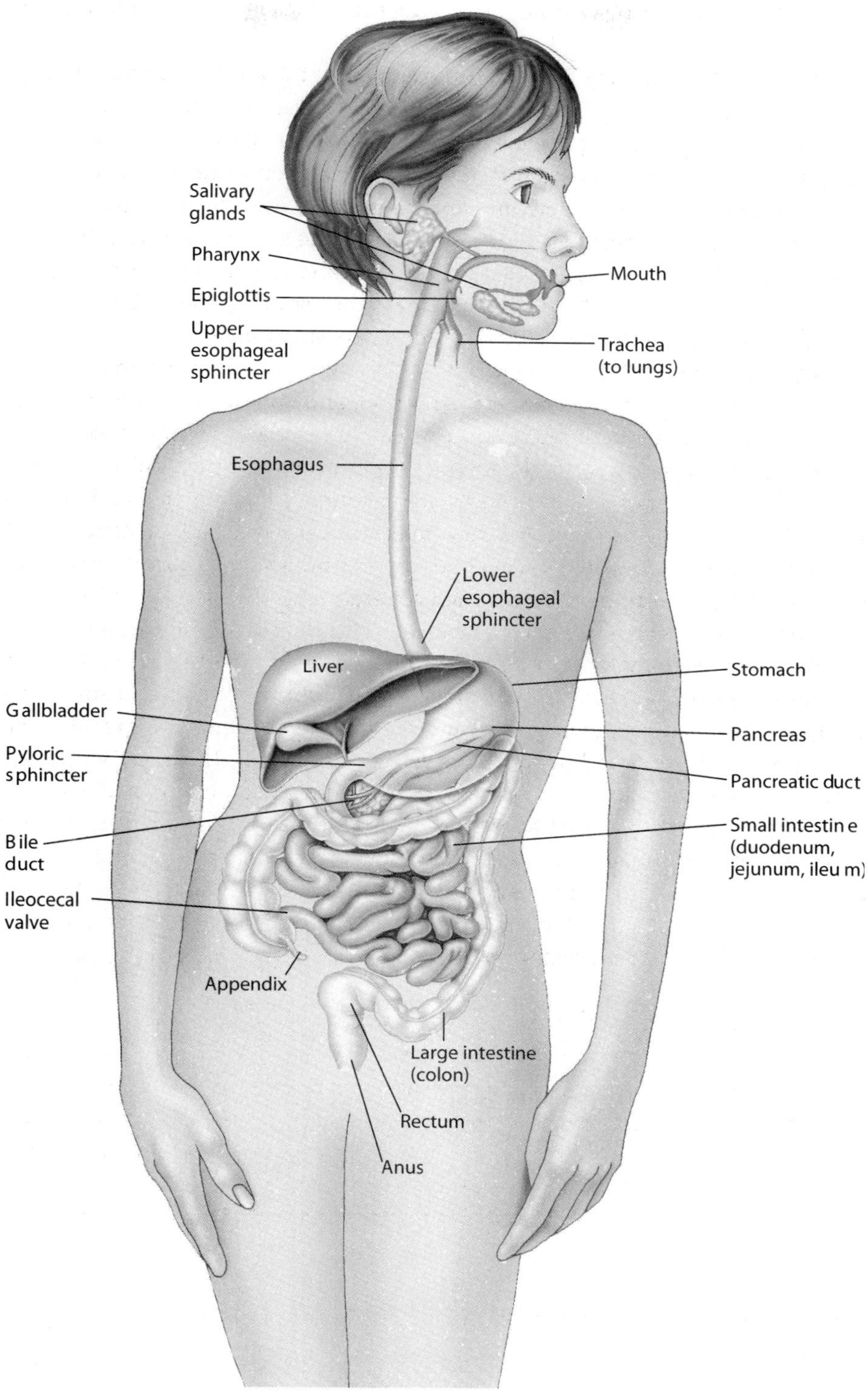
Salivary glands
Pharynx
Epiglottis
Upper esophageal sphincter
Mouth
Trachea (to lungs)
Esophagus
Lower esophageal sphincter
Liver
Stomach
Gallbladder
Pancreas
Pyloric sphincter
Pancreatic duct
Bile duct
Small intestine (duodenum, jejunum, ileum)
Ileocecal valve
Appendix
Large intestine (colon)
Rectum
Anus

	Fiber	Carbohydrate	Fat
Mouth	The mechanical action of the mouth crushes and tears fiber in food and mixes it with saliva to moisten it for swallowing.	The salivary glands secrete a watery fluid into the mouth to moisten the food. The salivary enzyme amylase begins digestion: starch $\xrightarrow{amylase}$ small polysaccharides, maltose	Glands in the base of the tongue secrete a lipase known as lingual lipase. Some hard fats begin to melt as they reach body temperature.
Esophagus	Fiber is unchanged.	Digestion of starch continues as swallowed food moves down esophagus.	Fat is unchanged.
Stomach	Fiber is unchanged.	Stomach acid and enzymes start to digest salivary enzymes, halting starch digestion. To a small extent stomach acid hydrolyzes maltose and sucrose: maltose $\xrightarrow{HCl}$ glucose sucrose $\xrightarrow{HCl}$ glucose & fructose	The lingual lipase hydrolyzes one bond of triglycerides to produce diglycerides and fatty acids. The degree of hydrolysis of fats by lingual lipase is slight for most fats but may be appreciable for milk fats.
Small intestine	Fiber is unchanged.	The pancreas produces carbohydrases and releases them through the pancreatic duct into the small intestine: polysaccharides $\xrightarrow{pancreatic\ amylase}$ maltose Then enzymes on the surfaces of the small intestinal cells break these into monosaccharides and the cells absorb them: maltose $\xrightarrow{maltase}$ sucrose $\xrightarrow{sucrase}$ glucose, fructose, galactose (absorbed) lactose $\xrightarrow{lactase}$	The stomach's churning action mixes fat with water and acid. A gastric lipase accesses and hydroylzes a very little fat. Bile flows in from the liver (via the common bile duct): fat $\xrightarrow{bile}$ emulsified fat Pancreatic lipase flows in from the pancreas: emulsified fat $\xrightarrow{pancreatic\ lipase}$ monoglycerides, glycerol, fatty acids (absorbed)

Fiber	Carbohydrate	Fat
Large intestine Most fiber passes intact through the digestive tract to the large intestine. Here, bacterial enzymes digest some fiber: some fiber $\xrightarrow{\textit{bacterial enzymes}}$ glucose (absorbed) Fiber holds water, regulates bowel activity, and binds cholesterol and some minerals, carrying them out of the body.	No action.	Some fat and cholesterol, trapped in fiber, exits in feces.

Protein	Vitamins	Minerals and Water
Mouth Chewing and crushing moistens protein-rich foods and mixes them with saliva to be swallowed.	No action on vitamins takes place in the mouth or esophagus.	The salivary glands add water to disperse and carry food.
Esophagus No action.	No action.	No action.
Stomach Stomach acid uncoils protein strands and activates stomach enzymes: protein $\xrightarrow[\text{HCl}]{\textit{pepsin}}$ smaller polypeptides	Intrinsic factor attaches to vitamin B_{12}.	Stomach acid (HCl) acts on iron to reduce it, making it more absorbable. The stomach secretes enough watery fluid to turn moist, chewed mass of solid food into liquid chyme.
Small intestine Pancreatic and small intestinal enzymes split polypeptides further: polypeptides $\xrightarrow{\textit{pancreatic and intestinal proteases}}$ dipeptides, tripeptides, and amino acids Then enzymes on the surface of the small intestinal cells hydrolyze these peptides and the cells absorb them: peptides $\xrightarrow{\textit{intestinal di- and tripeptidases}}$ amino acids (absorbed)	Bile emulsifies fat-soluble vitamins and aids in their absorption with other fats. Water-soluble vitamins are absorbed.	The small intestine, pancreas, and liver add enough fluid so that the total secreted into the intestine in a day approximates 2 gallons. Many minerals are absorbed. Vitamin D aids in the absorption of calcium.
Large intestine No action.	Bacteria produce vitamin K, which is absorbed.	More minerals and most of the water are absorbed.

2. Digestive Secretions and Their Actions

Organ or Gland	Secretion → Target	Action
salivary glands	*saliva* → mouth	starch —*amylase*→ **maltose**
gastric glands	*gastric juice* → **stomach**	protein —*pepsin*/HCl→ smaller polypeptides
intestinal glands	*intestinal juice* → small intestine	carbohydrate —*carbohydrase*→ monosaccharides
pancreas	*pancreatic juice* → small intestine	protein —*protease*→ dipeptides, tripeptides, **amino acids**
liver	*bile* → **gall bladder** —*bile*→ small intestine	fats —*bile*→ (emulsified fat) —*lipase*→ monoglycerides, **glycerol** and fatty acids

Sample Test Questions--

1. b (p. 77)
2. b (p. 76)
3. c (p. 76)
4. d (p. 79)
5. a (p. 81)
6. a (p. 80, 81)
7. d (p. 77)
8. d (p. 84)
9. a (p. 80)
10. b (p. 86)
11. c (p. 77)
12. a (p. 76)
13. d (p. 80)
14. e (p. 89)
15. d (p. 80)
16. c (p. 91)
17. b (p. 95)
18. c (p. 97)

Chapter 4
The Carbohydrates: Sugars, Starches, and Fibers

Chapter Outline

I. The Chemist's View of Carbohydrates

II. The Simple Carbohydrates
    A. Monosaccharides

    B. Disaccharides

III. The Complex Carbohydrates
    A. Glycogen

    B. Starches

    C. Fibers

IV. Digestion and Absorption of Carbohydrates
    A. Carbohydrate Digestion

    B.   Carbohydrate Absorption

    C.   Lactose Intolerance

V.   Glucose in the Body
    A.   A Preview of Carbohydrate Metabolism

    B.   The Constancy of Blood Glucose

VI.   Health Effects and Recommended Intakes of Sugars
    A.   Health Effects of Sugars

    B.   Accusations against Sugars

    C.   Recommended Intakes of Sugars

VII.   Health Effects and Recommended Intakes of Starch and Fibers
    A.   Health Effects of Starch and Fibers

    B.   Recommended Intakes of Starch and Fibers

C.   From Guidelines to Groceries

_____
_____
_____
_____

Highlight: Alternatives To Sugar
_____
_____
_____
_____
_____

SUMMING UP

At least half our food energy is derived from (1)_____, principally from (2)_____, but also from the simple sugars. Carbohydrates are classified as (3)_____ carbohydrates [the (4)_____] or simple carbohydrates [the (5)_____ saccharides and disaccharides]. Each of the three disaccharides [(6)_____, (7)_____, and maltose] contains a molecule of glucose paired with either (8)_____, galactose, or another glucose.

The polysaccharides starch and glycogen are composed of chains of (9)_____ units. (10)_____ is the storage form of glucose in the plant. Sources of starch in the diet include seeds, (11)_____, and starchy vegetables. (12)_____, or animal starch, is more complex than starch and is synthesized in (13)_____ and muscle from excess glucose in the bloodstream. The fibers include the polysaccharides (14)_____, pectin, and hemicellulose, mucilages, and gums as well as the nonpolysaccharide lignin.

Nutrition status affects our well-being even at the level of the body's (15)_____. The body strives to maintain its blood glucose within a normal range for optimal health and functioning. The hormones (16)_____, glucagon, and epinephrine function to maintain glucose (17)_____ in the body.

It is recommended that people consume (18)_____ to (19)_____ percent of the total kcalories from carbohydrates, preferably from (20)_____ carbohydrates. The (21)_____ system provides a useful guide to estimating the carbohydrate content of a meal.

Plant fibers also include the polysaccharides cellulose, (22)_____, and hemicellulose. There are also carbohydrate-like sources of fiber, such as (23)_____ and gums. Intakes of (24)_____ to (25)_____ grams of dietary fiber per day seem to be safe and beneficial.

ASSIGNMENTS

Answer these chapter study questions from the text:

1. Which carbohydrates are described as simple and which are complex?

2. Describe the structure of a monosaccharide and name the three monosaccharides important in nutrition. Name the three disaccharides commonly found in foods and their component monosaccharides. In what foods are these sugars found?

3. What happens in a condensation reaction? In a hydrolysis reaction?

4. Describe the structure of polysaccharides and name the ones important in nutrition. How are starch and glycogen similar and how do they differ? How do the fibers differ from the other polysaccharides?

5. Describe carbohydrate digestion and absorption. What role does fiber play in the process?

6. What are the possible fates of glucose in the body? What is the protein-sparing action of carbohydrate?

7. How does the body maintain blood glucose concentrations? What happens when it rises too high or falls too low?

8. What are the health effects of sugars? What are the dietary recommendations regarding concentrated sugar intakes?

9. What are the health effects of starches and fibers? What are the dietary recommendations regarding these complex carbohydrates?

10. What foods provide starches and fibers?

Chapter Glossary

acid-base balance: the equilibrium in the body between acid and base concentrations.
acidophilus milk: a cultured milk created by adding a bacterium that breaks down lactose to glucose and galactose.
amylase: an enzyme that hydrolyzes amylose (a form of starch).
brown sugar: refined white sugar crystals to which manufacturers have added molasses syrup with natural flavor and color; 91 to 96 percent pure sucrose.
carbohydrates: compounds composed of carbon, oxygen, and hydrogen arranged as monosaccharides or multiples of monosaccharides.
complex carbohydrates (starches and fibers): polysaccharides composed of straight or branched chains of monosaccharides.
condensation: a chemical reaction in which two reactants combine to yield a larger product.
confectioner's sugar: finely powdered sucrose; 99.9 percent pure.
corn sweeteners: corn syrup and sugars derived from corn.
corn syrup: a syrup produced by the action of enzymes on cornstarch; contains mostly glucose.
dental caries: decay of teeth.
dextrose: an older name for glucose.
diabetes: a disorder of carbohydrate metabolism resulting from inadequate or ineffective insulin.
disaccharide: a pair of monosaccharides linked together.
diverticula: a sac or pouch that develops in the weakened areas of the intestinal wall.
diverticulitis: the condition of having infected or inflamed diverticula.
diverticulosis: the condition of having diverticula.
empty-kcalorie food: a popular term used to denote foods that contribute energy but lack protein, vitamins, and minerals.

epinephrine: a hormone of the adrenal gland that modulates the stress response; formerly called *adrenaline*.
ferment: to digest in the absence of oxygen.
fiber: a general term denoting in plant foods the *nonstarch polysaccharides* that are not digested by *human* digestive enzymes.
fructose: a monosaccharide; sometimes known as fruit sugar or **levulose**, fructose is found abundantly in fruits, honey, and saps.
galactose: a monosaccharide; part of the disaccharide lactose.
glucagon: a hormone secreted by special cells in the pancreas in response to low blood glucose concentration that elicits release of glucose from storage.
gluconeogenesis: the making of glucose from a noncarbohydrate source.
glucose: a monosaccharide; sometimes known as blood sugar or *dextrose*.
glycemic effect: a measure of the extent to which food, as compared with pure glucose, raises the blood-glucose concentration and elicits an insulin response.
glycemic index: a method used to classify foods according to their potential for raising insulin response.
glycogen: an animal polysaccharide composed of glucose; it is manufactured and stored in the liver and muscles as a storage form of glucose.
granulated sugar: crystalline sucrose; 99.9 percent pure.
hexoses: the monosaccharides important in nutrition, they have the formula $C_6H_{12}O_6$.
high-fructose corn syrup (HFCS): a corn-syrup sweetener made especially for use in processed foods and beverages, where it is the predominant sweetener.
honey: sugar (mostly sucrose) formed from nectar gathered by bees. An enzyme splits the sucrose into glucose and fructose. Composition and flavor vary, but honey always contains a mixture of sucrose, fructose, and glucose.
hypoglycemia: an abnormally low blood glucose concentration.
insoluble fibers: indigestible food components that do not dissolve in water.
insulin: a hormone secreted by special cells in the pancreas in response to (among other things) increased blood glucose concentration.
insulin-dependent diabetes mellitus (IDDM): the less common type of diabetes in which the person produces no insulin at all; also known as *type I diabetes* or *juvenile-onset diabetes*.
invert sugar: a mixture of glucose and fructose formed by the hydrolysis of sucrose in a chemical process; sold only in liquid form and sweeter than sucrose.
ketone bodies: the product of the incomplete breakdown of fat when glucose is not available in the cells.
ketosis: an undesirably high concentration of ketone bodies in the blood and urine.
lactase: an enzyme that hydrolyzes lactose.
lactase deficiency: a lack of the enzyme required to digest the disaccharide lactose into its component monosaccharides (glucose and galactose).
lactose: a disaccharide composed of glucose and galactose; commonly known as milk sugar.
lactose intolerance: a condition that results from inability to digest the milk sugar lactose; characterized by bloating, gas, abdominal discomfort, and diarrhea.
levulose: an older name for fructose.
maltase: an enzyme that hydrolyzes maltose.
maltose: a disaccharide composed of two glucose units; sometimes known as malt sugar.
maple sugar: a sugar (mostly sucrose) purified from the concentrated sap of the sugar maple tree.
molasses: the thick brown syrup produced during sugar refining.
monosaccharide: a carbohydrate of the general formula $C_nH_{2n}O_n$ that consists of a single ring.
noninsulin-dependent diabetes mellitus (NIDDM): the more common type of diabetes in which the fat cells resist insulin; also called *type II diabetes* or *adult-onset diabetes*.

phytic acid: a nonnutrient component of plant sees; also called *phytate*.
plaque (dental): a gummy mass of bacteria that grows on teeth and can lead to dental caries and gum disease.
polysaccharide: many monosaccharides linked together.
protein-sparing action: the action of carbohydrate (and fat) in providing energy that allows protein to be used for other purposes.
raw sugar: the first crop of crystals harvested during sugar processing.
resistant starch: starch that escapes digestion and absorption in the small intestine of healthy people.
satiety: the feeling of fullness and satisfaction that food brings.
simple carbohydrates (sugars): monosaccharides and disaccharides.
soluble fibers: indigestible food components that dissolve in water to form a gel.
starches: plant polysaccharides composed of glucose that are digestible by human beings.
sucrase: an enzyme that hydrolyzes sucrose.
sucrose: a disaccharide composed of glucose and fructose; commonly known as table sugar, beet sugar, or cane sugar.
turbinado sugar: sugar produced using the same refining process as white sugar, but without the bleaching and anti-caking treatment; traces of molasses give turbinado its sandy color.
type I diabetes: the less common type of diabetes in which the person produces no insulin at all.
type II diabetes: the more common type of diabetes in which the fat cells resist insulin.
white sugar: pure sucrose or "table sugar," produced by dissolving, concentrating, and recrystallizing raw sugar.

Complete these short answer questions:

1. Hormones involved in maintaining blood glucose are:

    a.   c.

    b.

2. The monosaccharides are:

    a.   c.

    b.

3. The disaccharides are:

    a.   c.

    b.

4. The polysaccharides are:

    a.   c.

    b.

Complete this crossword puzzle by Mary A. Wyandt, Ph.D., CHES.

Across	Down
3. an enzyme that hydrolyzes amylose	1. a non-nutrient component of plant seeds; also called phytate
5. to digest in the absence of oxygen	
6. an abnormally low blood glucose concentration	2. a sac or pouch that develops in the weakened areas of the intestinal wall
8. an enzyme that hydrolyzes maltose	
9. a disorder of carbohydrate metabolism resulting from inadequate or ineffective insulin	4. an undesirably high concentration of ketone bodies in the blood and urine
10. a hormone secreted by special cells in the pancreas in response to low blood glucose concentration that elicits release of glucose from storage	7. a hormone secreted by special cells in the pancreas in response to (among other things) increased blood glucose concentration

Identify these structures and reactions:

1. 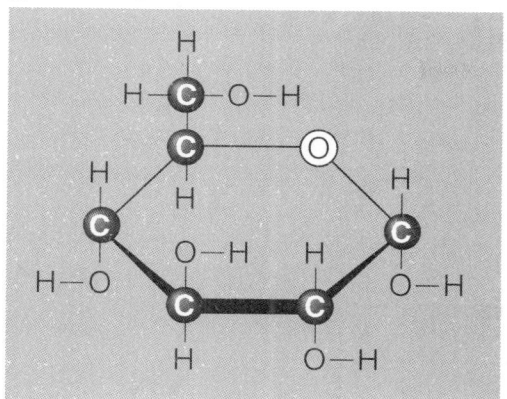
2. 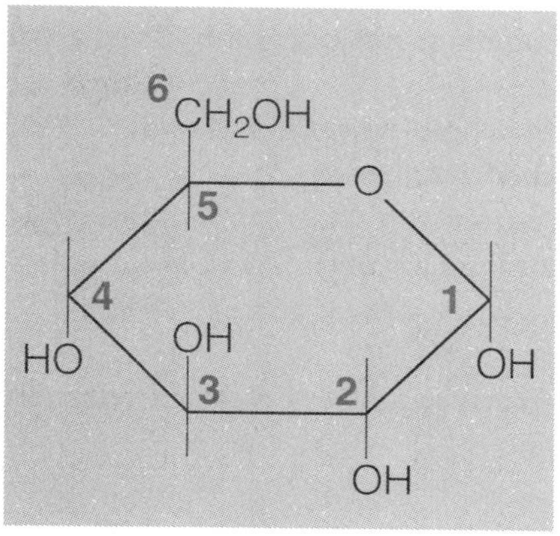
3. 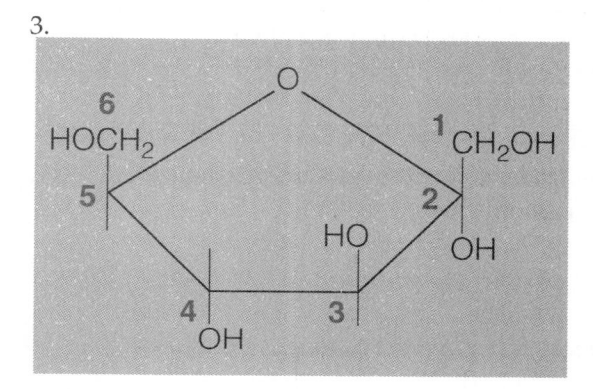
4. 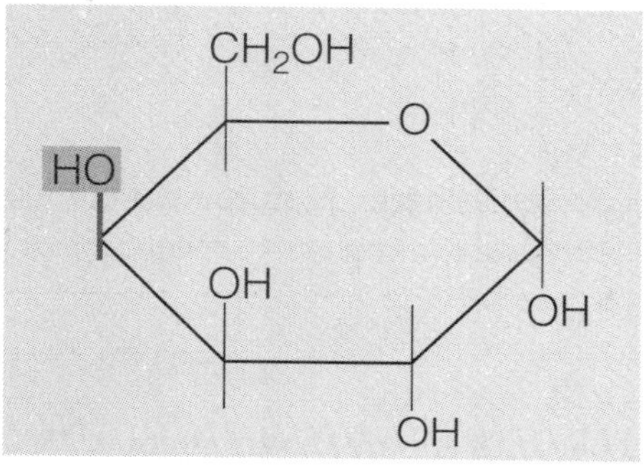
5. 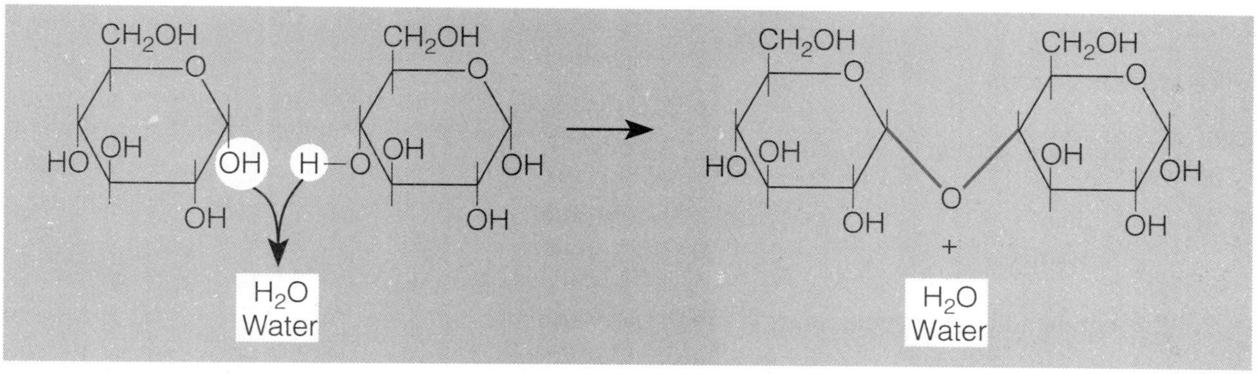
6.

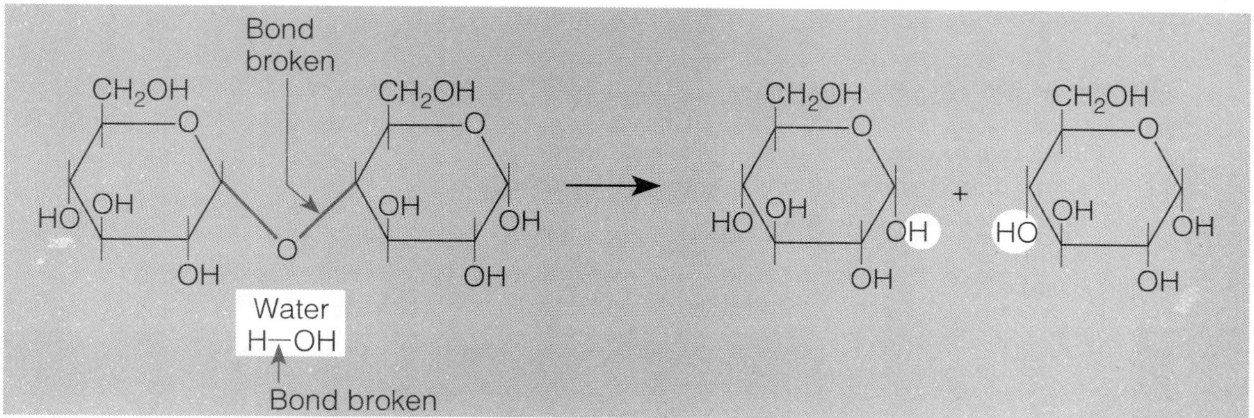

Solve these problems:

1. How many grams of carbohydrate are in the following meal?

    2 slices whole-wheat bread with
    1 oz cheddar cheese
    1 pat butter
    1/2 c orange juice

2. How many kcalories do carbohydrates contribute to the meal above?

3. The meal provides 365 kcalories. What percentage of the kcalories is from carbohydrates?

4. If a person's energy requirement is 1500 kcalories per day, calculate the number of grams of carbohydrate necessary to meet 55% of kcalories from carbohydrate?

5. If a person's energy requirement is 2500 kcalories per day, calculate the number of grams of carbohydrate necessary to meet 55% of kcalories from carbohydrate.

SAMPLE TEST QUESTIONS

1. Carbohydrates appear in virtually all _____ foods.

    a. plant   c. health
    b. animal   d. protein

2. Which of the following compounds is a monosaccharide?

    a. sucrose   d. lactose
    b. fructose   e. pectin
    c. maltose

3. A hydrolysis reaction can:

    a. bond two monosaccharides to form a disaccharide.
    b. split a disaccharide to form two monosaccharides.
    c. form a molecule of water.
    d. a and c.
    e. b and c.

4. This reaction links two monosaccharides together:

    a. disaccharide.
    b. hydrolysis.
    c. absorption.
    d. condensation.

5. Disaccharides include:

    a. glucose.
    b. maltose.
    c. glycogen.
    d. sucrose.
    e. b and d.

6. The principal carbohydrate of milk is:

    a. lactose.
    b. sucrose.
    c. maltose.
    d. glycogen.

7. Fruits are usually sweet because they contain:

    a. fiber.
    b. complex carbohydrates.
    c. simple sugars.
    d. fats.

8. An animal polysaccharide composed of glucose is called:

    a. fiber.
    b. dextrins.
    c. glycogen.
    d. an enzyme.

9. The difference between glycogen and starch is:

    a. the bonds in starch are fatty acids.
    b. the bonds in starch are single.
    c. the bonds in glycogen are not hydrolyzed by human enzymes.
    d. the glucose units are linked together differently.

10. Starch is made up of many glucose units bonded together.

    a. true
    b. false

11. Most carbohydrate absorption occurs in the:

    a. mouth.
    b. stomach.
    c. small intestine.
    d. large intestine.
    e. b and c.

12. An enzyme that hydrolyzes sucrose is:

    a. glucose.
    b. lactase.
    c. pectins.
    d. sucrase.

13. Carbohydrate digestion occurs in:

    a. the stomach and small intestine.
    b. the mouth and small intestine.
    c. the stomach and colon.
    d. the mouth and pancreas.

14. A condition that results from inability to digest lactose is called:

    a. lactose deficiency.
    b. lactose intolerance.
    c. hypoglycemia.
    d. lactase intolerance.

15. The main function of carbohydrate in the body is to:

    a. furnish the body with energy.
    b. provide materials for synthesizing cell walls.
    c. synthesize fat.
    d. insulate the body to prevent heat loss.

16. When blood glucose levels fall, the liver:

    a. combines excess glucose molecules.
    b. stores glucose as glycogen.
    c. dismantles stored glycogen.
    d. combines glucose to form molecules of fat.

17. Sugar causes this (these) harmful condition(s):

    a. addictiveness.
    b. diabetes.
    c. ulcers.
    d. dental caries.
    e. all of the above.

18. Taken with ample fluids, fibers can help prevent the following disorder(s):

    a. hemorrhoids.
    b. appendicitis.
    c. hyperactivity.
    d. a and b.
    e. b and c.

19. Dietary fiber:

    a. raises blood cholesterol levels.
    b. is found in high fat foods.
    c. causes diverticulosis.
    d. provides satiety and delays hunger.

20. Excess fiber can result in:

   a. abdominal discomfort.
   b. gas and diarrhea.
   c. dental caries.
   d. a and b.
   e. a, b, and c.

ANSWERS

Summing Up—(1) carbohydrate; (2) starch; (3) complex; (4) polysaccharides; (5) mono; (6) sucrose; (7) lactose; (8) fructose; (9) glucose; (10) starch; (11) grains; (12) glycogen; (13) liver; (14) cellulose; (15) cells; (16) insulin; (17) homeostasis; (18) 55; (19) 60; (20) complex; (21) exchange; (22) pectin; (23) lignin; (24) 20; (25) 35.

Chapter study questions from the text—(1) Simple—monosaccharides (glucose, fructose, galactose), disaccharides (sucrose, lactose, maltose). Complex—glycogen and starch. (2) A carbohydrate with a general structure of a ring composed of carbon, hydrogen and oxygen atoms. Monosaccharides important in nutrition are glucose, fructose, and galactose; disaccharides important in nutrition are sucrose (fructose + glucose), lactose (galactose + glucose), and maltose (2 glucose). Nearly all plant foods contain glucose; most plants (especially fruits and saps) contain fructose; galactose is not found as such in foods; sucrose occurs in many fruits and some vegetables and grains; lactose is found in milk; maltose is found in seeds. (3) Condensation combines two reactants to yield one product with the removal of water; hydrolysis splits one reactant into two products with the addition of water. (4) Polysaccharides are composed of many monosaccharides strung together. Important in nutrition are: glycogen, starch and the fibers. Starch and glycogen are similar in that they are both composed of glucose; they differ in the way their glucose units are linked together (glycogen consists of many glucose molecules linked together in highly branched chains, while starch consists of many glucose molecules linked side by side). Fibers are different in that the bonds between their monosaccharides cannot be broken. (5) Salivary amylase enzymes in the mouth partially break down some of the starch before it reaches the intestine, pancreatic enzymes digest the starch to disaccharides in small intestine, disaccharidase enzymes on surface of intestinal wall cells split disaccharides to monosaccharides, monosaccharides enter capillary, capillary delivers monosaccharides to liver, liver converts galactose and fructose to glucose. In the mouth, fiber slows the process of eating and stimulates the flow of saliva; in the stomach, they delay gastric emptying; in the small intestine, they delay absorption of carbohydrates and fats, and can bind with minerals; in the large intestine, they attract water that softens the stools. (6) It can be stored as glycogen; it can be used for energy; it can be converted to fat, when carbohydrate is available; it can be used for energy, leaving protein available for its special functions. (7) Hormones are secreted in response to fluctuations in blood glucose. When blood glucose is too high, the pancreas releases insulin, resulting in the storage of glucose in the cells; when blood glucose is too low, the pancreas releases glucagon, resulting in the release of glucose into the blood. (8) Excess sugar can cause malnutrition if sugar displaces needed nutrients from the diet. It can contribute to obesity. Excess sugar can contribute to elevated blood lipids, and it can cause dental caries. Less than 10% of total kcalories should come from concentrated sugar. (9) Protect against heart disease, colon cancer, and diabetes, assist in weight control, improve large intestine function and health, lower blood cholesterol levels, and slow the rate of glucose absorption. 55 to 60% of total kcalories should come from carbohydrate; mostly from starch, some from fruits, vegetables, and milk. (10) Fruits, vegetables, legumes, whole-grain breads and cereals.

Short Answers—
1. insulin; glucagon; epinephrine
2. glucose; fructose; galactose
3. maltose; sucrose; lactose
4. starch; glycogen; cellulose

Identification of structures and reactions—
1. glucose
2. glucose
3. fructose
4. galactose
5. condensation reaction
6. hydrolysis reaction

Problem Solving—
1. 2 (15 g) + 0 + 0 + 15 g = 45 g
2. 4 kcal/g X 45 g = 180 kcal
3. 180 kcal divided by 365 kcal = 49%
4. 1500 kcal X .55 = 825 kcal divided by 4 = 206 grams
5. 2500 kcal X .55 = 1375 kcal divided by 4 = 344 grams

Sample Test Questions—
1. a (p. 103)
2. b (p. 104)
3. b (p. 106)
4. d (p. 106)
5. e (p. 105)
6. a (p. 106)
7. c (p. 106, 107)
8. c (p. 108)
9. d (p. 108)
10. a (p. 108)
11. c (p. 111)
12. d (p. 111)
13. b (p. 111)
14. b (p. 113)
15. a (p. 114)
16. c (p. 115)
17. d (p. 121)
18. d (p. 126)
19. d (p. 125)
20. d (p. 126)

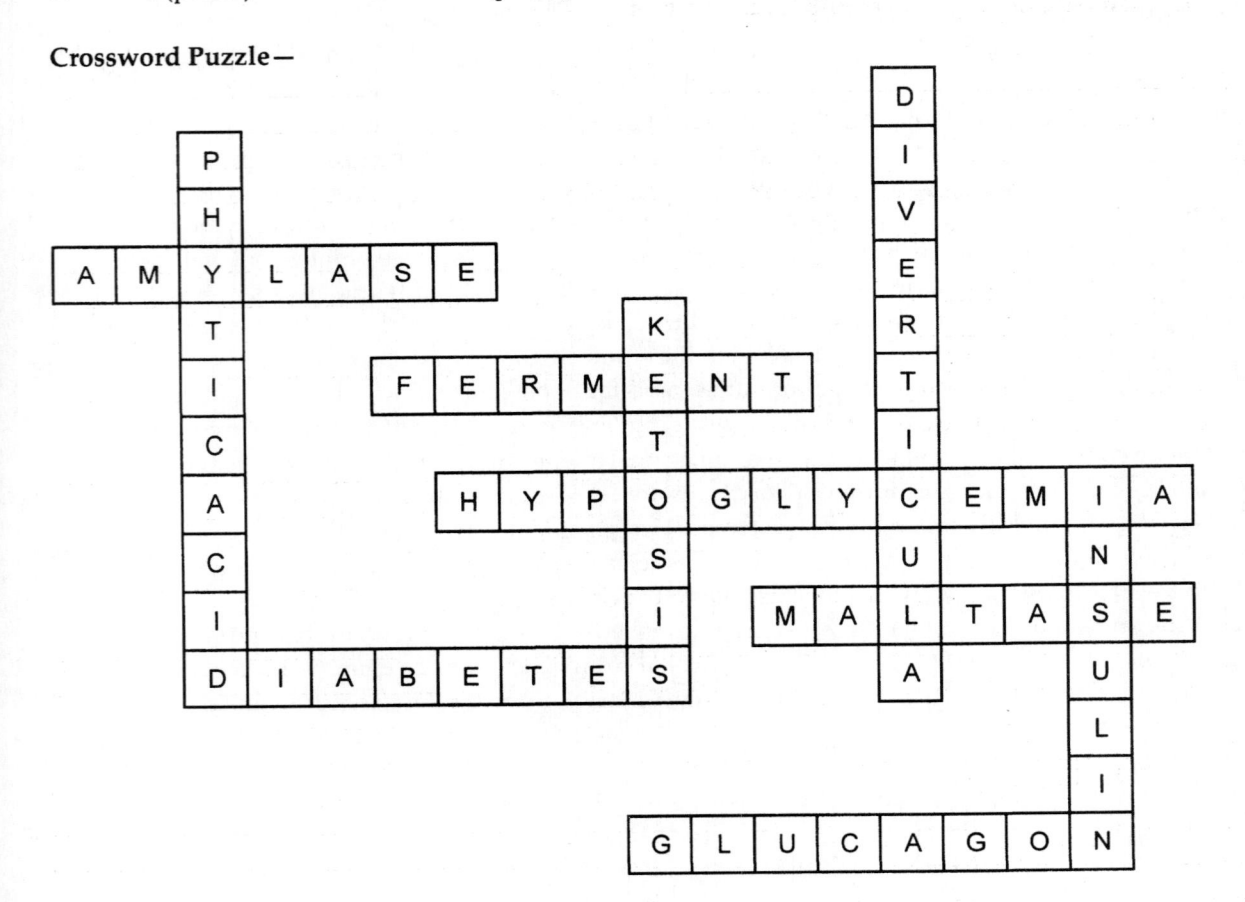

Crossword Puzzle—

CHAPTER 5
THE LIPIDS: TRIGLYCERIDES, PHOSPHOLIPIDS, AND STEROLS

CHAPTER OUTLINE

I. The Chemist's View of Fatty Acids and Triglycerides
    A. Fatty Acids

    B. Triglycerides

    C. Degree of Unsaturation Revisited

II. The Chemist's View of Phospholipids and Sterols
    A. Phospholipids

    B. Sterols

III. Digestion, Absorption, and Transport of Lipids
    A. Lipid Digestion

    B. Lipid Absorption

    C. Lipid Transport

IV. Lipids in the Body
   A. Roles of Triglycerides

   B. Essential Fatty Acids

   C. A Preview of Lipid Metabolism

V. Health Effects and Recommended Intakes of Lipids
   A. Health Effects of Lipids

   B. Recommended Intakes of Fat

   C. From Guidelines to Groceries

Highlight: High-Fat Foods—Friend or Foe?

SUMMING UP

Lipids in the body, including (1)_____, oils, phospholipids, and sterols, function to maintain the health of the (2)_____ and hair; to protect body (3)_____ from heat, cold, and mechanical shock; and to provide a continuous (4)_____ supply. In foods, fats and oils act as a solvent for the fat-soluble (5)_____ and the compounds that give foods their (6)_____ and aromas.

About 95 percent of the lipids in the diet are (7)_____; the phospholipids and sterols make up the remaining 5 percent. Triglycerides are composed of (8)_____ with (9)_____ fatty acids attached. The fatty acids may be classified as (10)_____, monounsaturated, or polyunsaturated.

During digestion, the triglycerides are emulsified by (11)_____ and then hydrolyzed by enzymes to (12)_____, glycerol, and fatty acids, which then pass into the intestinal cells. After absorption, all three classes of lipids are transported by (13)_____ in the body fluids. Cholesterol, a sterol, is synthesized in the body by the (14)_____. The body uses cholesterol in cell membranes and to make (15)_____ salts, hormones, and vitamin D.

Cholesterol from the liver may be transported to body tissues via the lipoproteins and may also be abnormally deposited in (16)_____ walls. A diet high in (17)_____ fat and cholesterol has been implicated as a causative factor in (18)_____. Most authorities recommend limiting excess fat, saturated fat, and cholesterol.

Most of the saturated fat found in the diet comes from (19)_____ and animal fats. Cholesterol is contributed by organ meats, shellfish, (20)_____, meats, and animal fats. (21)_____ products do not contain cholesterol. Vegetable and fish oils generally contain more (22)_____ fats than do animal fats.

Assignments

Answer these chapter study questions from the text:

1. Name the three classes of lipids found in the body and in foods. What are some of their functions in the body? What features do fats bring to foods?

2. What features distinguish fatty acids from each other?

3. What does the term "omega" mean with respect to fatty acids? Describe the roles of the omega fatty acids in disease prevention.

4. What are the differences between saturated, unsaturated, monounsaturated, and polyunsaturated fats? Describe the structure of a triglyceride.

5. What does hydrogenation do to fat? What are *trans*-fatty acids and how do they influence heart disease?

6. How do phospholipids differ from triglycerides in structure? How does cholesterol differ? How do these differences in structure affect function?

7. What roles do phospholipids play in the body? What roles does cholesterol play in the body?

8. Trace the steps in fat digestion, absorption, and transport. Describe the routes cholesterol takes in the body.

9. What do lipoproteins do? What are the differences among the chylomicrons, VLDL, LDL, and HDL?

10. Which of the fatty acids are essential? Name their chief dietary sources.

11. How does excessive fat intake influence health? What factors influence LDL, HDL, and total blood cholesterol?

12. What are the dietary recommendations regarding fat and cholesterol intake? List ways to reduce intake.

13. What is the Daily Value for fat (for a 2000-kcalorie diet)? What does this number represent?

CHAPTER GLOSSARY

acid group: the COOH group of an organic acid.
adipose tissue: the body's fat tissue, which consists of masses of fat-storing cells.
antioxidants: compounds that protect others from oxidation by being oxidized themselves.
arachidonic acid: an omega-6 polyunsaturated fatty acid with 20 carbons and four double bonds (20:4); synthesized from linoleic acid.
artificial fats: zero-energy fat replacers that are chemically synthesized to mimic the sensory and cooking qualities of naturally occurring fats, but are totally or partially resistant to digestion.
atherosclerosis: a type of artery disease characterized by accumulations of lipid-containing material on the inner walls of the arteries.
blood lipid profile: the results of blood tests that reveal a person's total cholesterol, triglycerides, and various lipoproteins.
cardiovascular disease (CVD): a general term for all diseases of the heart and blood vessels.
cholesterol: one of the sterols containing four carbon rings and a carbon side chain.
choline: a nitrogen-containing compound found in plant and animal tissues as part of lecithin and other phospholipids.
chylomicrons: the class of lipoproteins that transport lipids from the intestinal cells into the body.
docosahexaenoic acid (DHA): an omega-3 polyunsaturated fatty acid with 22 carbons and six double bonds (22:6); synthesized from linolenic acid.
eicosanoids: derivatives of fatty acids; hormonelike compounds that regulate blood pressure, clotting, and other body functions.
eicosapentaenoic acid (EPA): an omega-3 polyunsaturated fatty acid with 20 carbons and five double bonds (20:5); synthesized from linolenic acid.
endogenous: made in the body.
enterohepatic circulation: process by which bile is released into the small intestine and reabsorbed and sent back to the liver to be reused.
essential fatty acids: fatty acids needed by the body, but not made by the body in amounts sufficient to meet physiological needs.
exogenous: obtained from outside the body.

fat replacers: ingredients that replace some or all of the functions of fat and may or may not provide energy.
fats: the lipids in foods or body fat, composed mostly of triglycerides.
fatty acid: an organic compound composed of a carbon chain with hydrogens attached and an acid group at one end.
glycerol: an alcohol composed of a three-carbon chain, which can serve as the backbone for a triglyceride.
HDL (high-density lipoprotein): the type of lipoprotein that transports cholesterol back to the liver from peripheral cells; composed primarily of protein.
hormone-sensitive lipase: an enzyme inside adipose cells that responds to the body's need for fuel by hydrolyzing triglycerides so that their parts (glycerol and fatty acids) escape into the general circulation and thus become available to other cells as fuel.
hydrogenation: a chemical process by which hydrogens are added to unsaturated or polyunsaturated fats to reduce the number of double bonds, making the fats more saturated and more resistant to oxidation.
hydrophilic: a term referring to water-loving, or water-soluble, substances.
hydrophobic: a term referring to water-fearing, or non-water-soluble, substances; also known as *lipophilic* (fat loving).
hypertension: high blood pressure.
LDL (low-density lipoprotein): the type of lipoprotein derived from very-low-density lipoproteins (VLDL) as cells remove triglycerides from them; composed primarily of cholesterol.
lecithin: one of the phospholipids; a compound of glycerol to which are attached two fatty acids, a phosphate group, and a choline molecule.
linoleic acid: an essential fatty acid with 18 carbons and two double bonds (18:2).
linolenic acid: an essential fatty acid with 18 carbons and three double bonds (18:3).
lipid profile: the results of blood tests that reveal a person's total cholesterol and triglycerides and the amounts of cholesterol in the various lipoproteins.
lipids: a family of compounds that includes triglycerides (fats and oils), phopholipids, and sterols.
lipoprotein lipase (LPL): an enzyme mounted on the surface of fat cells (and other cells) that hydrolyzes triglycerides passing by in the bloodstream and directs their parts into the cells, where they can be metabolized or reassembled for storage.
lipoproteins: clusters of lipids associated with proteins that serve as transport vehicles for lipids in the lymph and blood.
micelles: tiny spherical complexes that arise during fat digestion, each carrying about 20 fatty acids and/or monoglycerides into intestinal cells.
monoglyceride: a molecule of glycerol with one fatty acid attached. A molecule of glycerol with two fatty acids attached is a *diglyceride*.
monounsaturated fatty acid: a fatty acid that lacks two hydrogen atoms and has one double bond between carbons--for example, oleic acid.
oils: liquid fats (at room temperature).
olestra: a synthetic fat made from sucrose and fatty acids that provides 0 kcalories per gram; also known as sucrose polymer.
omega: the last letter of the Greek alphabet used by chemists to refer to the position of the endmost double bond in a fatty acid.
omega-3 fatty acid: a polyunsaturated fatty acid in which the first double bond is three carbons away from the methyl (CH_3) end of the carbon chain.
omega-6 fatty acid: a polyunsaturated fatty acid in which the first double bond is six carbons from the methyl (CH_3) end of the carbon chain.

oxidation: the process of a substance combining with oxygen.
palatability: pleasing taste.
phospholipid: a compound similar to a triglyceride but having choline (or another nitrogen-containing compound) and a phosphate group (a phosphorus-containing salt) in place of one of the fatty acids.
point of unsaturation: the double bond of a fatty acid, where hydrogen atoms can easily be added to the structure.
polyunsaturated fatty acid (PUFA): a fatty acid that lacks four or more hydrogen atoms and has two or more double bonds between carbons—for example, linoleic acid (two double bonds) and linolenic acid (three double bonds).
saturated fat: composed of triglycerides in which all or virtually all of the fatty acids are saturated.
saturated fatty acid: a fatty acid carrying the maximum possible number of hydrogen atoms--for example, stearic acid.
sterols: compounds composed of C, H, and O atoms arranged in rings like those of cholesterol, with any of a variety of side chains attached.
***trans*-fatty acids:** fatty acids with an unusual configuration around the double bond.
triglycerides: the chief form of fat in the diet and the major storage form of fat in the body; composed of a molecule of glycerol with three fatty acids attached, also called *triacylglycerols*.
unsaturated fatty acid: a fatty acid that lacks hydrogen atoms and has at least one double bond between carbons (includes monounsaturated and polyunsaturated fatty acids).
VLDL (very-low-density lipoprotein): the type of lipoprotein made primarily by liver cells to transport lipids to various tissues in the body; composed primarily of triglycerides.

Complete these short answer questions:

1. The lipids include:

    a.

    b.

    c.

2. The roles of body fat are:

    a.   c.

    b.   d.

3. The roles of food fat are:

    a.   c.

    b.   d.

4. Triglycerides are made of:

    a.   b.

5. The lipoproteins are:

   a.   c.

   b.   d.

Identify the following chemical structures:

1.

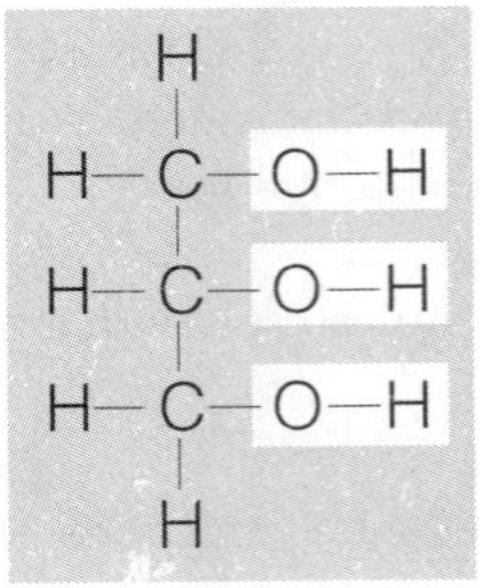

2.

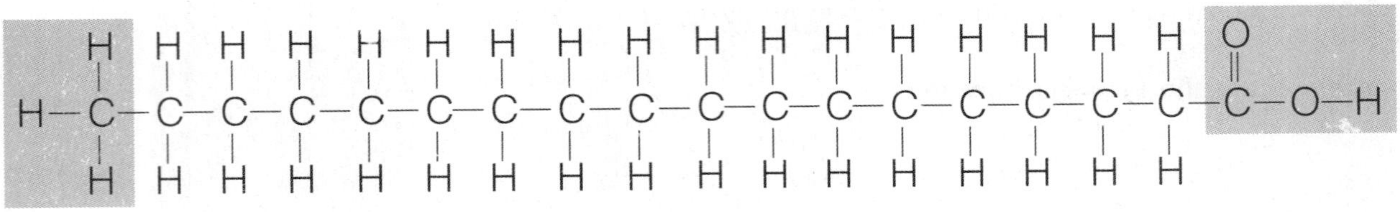

3.

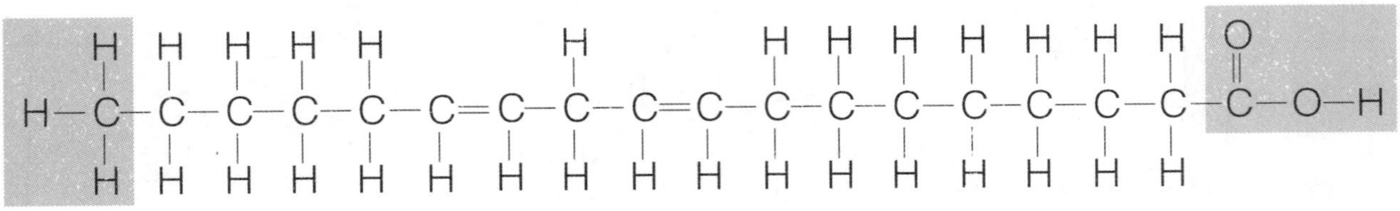

4.

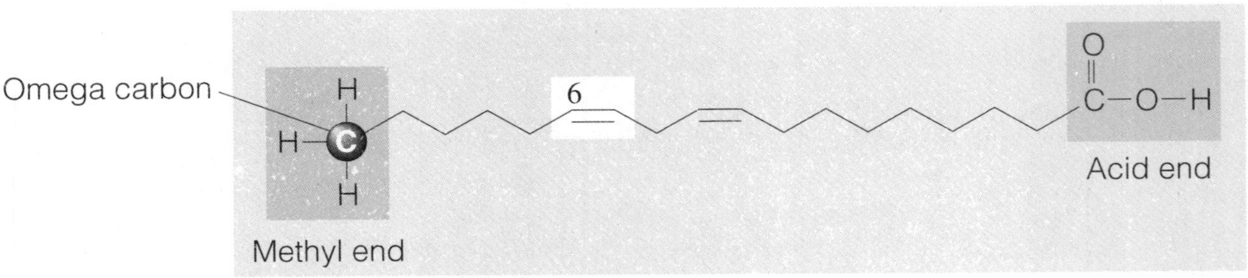

Omega carbon

Methyl end

6

Acid end

5.

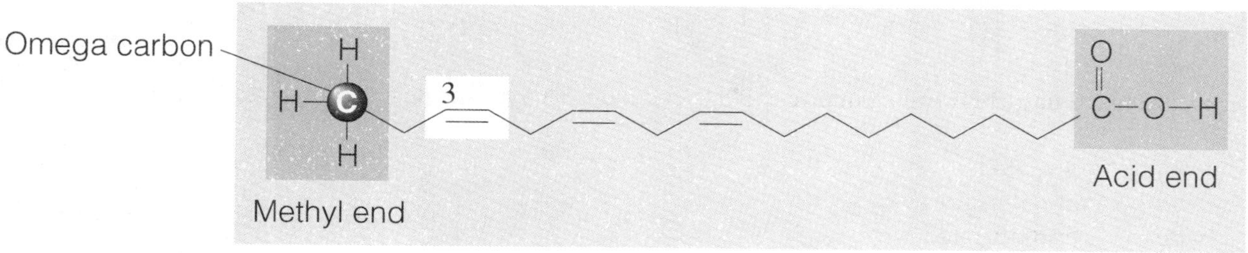
Omega carbon
H
H
C
H
3
O
C–O–H
Methyl end
Acid end

6.

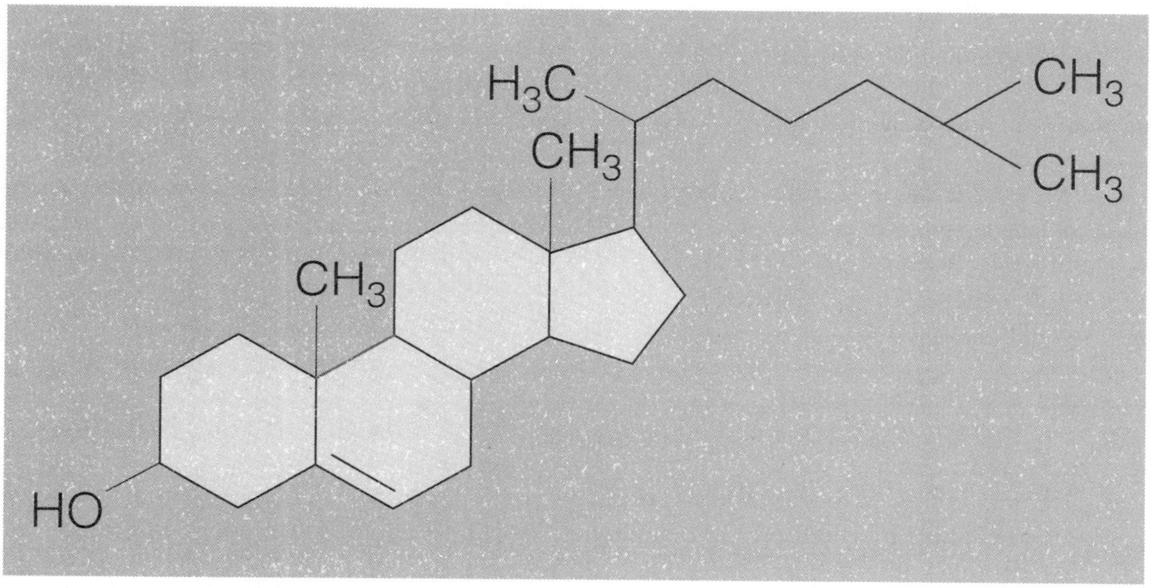
H3C
CH3
CH3
CH3
CH3
HO

SAMPLE TEST QUESTIONS

1. Fats belong to a larger chemical classification known as:

   a. lipases.
   b. lecithins.
   c. labiles.
   d. leucines.
   e. lipids.

2. Of the lipids in foods, 95% are triglycerides.

   a. true
   b. false

3. Which of the following is not lost when fat or oil is removed from a food?

   a. flavor
   b. aroma
   c. kcalories
   d. vitamin E
   e. vitamin C

4. Triglycerides are composed of:

   a. glycerol.
   b. pyruvate.
   c. fatty acids.
   d. a and c.
   e. b and c.

5. Oleic acid has one double bond in the carbon chain. This means it is classified as:

   a. saturated.
   b. monounsaturated.
   c. diunsaturated.
   d. polyunsaturated.

6. Linoleic acid is classified as an essential fatty acid because it:

   a. neutralizes cholesterol.
   b. cannot be made in the body.
   c. is a polyunsaturated fatty acid.
   d. is found in animal fats.

7. A disadvantage of hydrogenation is:

   a. the fats become resistant to oxidation.
   b. cis-fatty acids result.
   c. trans-fatty acids result.
   d. a and b.
   e. b and c.

8. Hydrogenation of fat renders it:

   a. more nutritious.
   b. more susceptible to oxidation and rancidity.
   c. less susceptible to oxidation and rancidity.
   d. less saturated, thus less stable.

9. The manufacturing process that makes polyunsaturated fats more solid is called:

   a. centrifugation.
   b. hydrogenation.
   c. emulsification.
   d. modification.
   e. solidification.

10. The dispersion and stabilization of fat droplets in a watery solution is:

    a. hydrogenation.
    b. emulsification.
    c. saturation.
    d. precipitation.

11. Lecithins are:

    a. carbohydrates.
    b. triglycerides.
    c. proteins.
    d. phospholipids.

12. Bile is made from:

    a. glucose.
    b. cholesterol.
    c. vitamin A.
    d. prostaglandins.

13. A lipoprotein may be described as:

    a. a molecule made up of various amino acids.
    b. a triglyceride made of glycerol and three fatty acids.
    c. a cluster of lipids wrapped in a coat of protein.
    d. undigested fat circulating in the bloodstream.

14. The largest and least dense of the lipoproteins are the:

    a. chylomicrons.
    b. LDL.
    c. VLDL.
    d. HDL.

15. The lipoprotein thought to protect against coronary heart disease is:

    a. LDL.
    b. VLD.
    c. IDL.
    d. HDL.
    e. chylomicrons.

16. Which of the following foods contains an appreciable amount of fat?

    a. skim milk
    b. bread
    c. meat
    d. vegetables
    e. fruits

17. Which contains the most saturated fat?

    a. sunflower seed oil
    b. corn oil
    c. lard
    d. olive oil

18. Which of the following foods contains no cholesterol?

    a. beef steak
    b. bacon
    c. ice cream
    d. baked potato

19. A change from whole milk to nonfat milk would:

    a. decrease the amount of fat and kcalories.
    b. increase the number of kcalories.
    c. increase satiety value of the meal.
    d. decrease the amount of essential amino acids.

Complete this crossword puzzle by Mary A. Wyandt, Ph.D., CHES.

Across:	Down:
1. One of the sterols containing a four-carbon ring structure with a carbon side chain.	2. A term referring to water-loving, or water-soluble, substances.
5. Results of blood tests that reveal a person's total cholesterol, triglycerides, and various lipoproteins.	3. The chief form of fat in the diet and the major storage form of fat in the body; composed of a molecule of glycerol with three fatty acids attached.
7. An alcohol composed of a three-carbon chain, which can serve as the backbone for a triglyceride.	4. A compound similar to a triglyceride but having a phosphate group and choline in place of one of the fatty acids.
8. High blood pressure.	6. Derivatives of 20-carbon fatty acids; biologically active compounds that help to regulate blood pressure, blood clotting, and other body functions.
9. Clusters of lipids associated with proteins that serve as transport vehicles for lipids in the lymph and blood.	10. The last letter of the Greek alphabet, used by chemists to refer to the position of the first double bond from the methyl end of a fatty acid.

Answers

Summing Up—(1) fats; (2) skin; (3) organs; (4) fuel; (5) vitamins; (6) flavor; (7) triglycerides; (8) glycerol; (9) 3; (10) saturated; (11) bile; (12) monoglycerides; (13) lipoproteins; (14) liver; (15) bile; (16) artery; (17) saturated; (18) atherosclerosis; (19) meat; (20) eggs; (21) vegetable; (22) polyunsaturated.

Chapter study questions from the text—(1) Triglycerides, phospholipids, sterols. Features: fats enhance foods' aroma and flavor, increase palatability, provide kcalories and fat-soluble vitamins. Functions: carry fat-soluble vitamins, induce satiety, provide body with a continuous food supply, keep body warm, protect it from mechanical shock, serve as starting materials for hormonal regulations; phospholipids and sterols contribute to cells' structures, cholesterol serves as raw material for hormones, vitamin D, and bile. (2) Essentiality and nonessentiality; size indicated by number of carbons; saturated versus unsaturated. (3) In polyunsaturated fatty acids, omega refers to the relative place in which the first double bond is located from the methyl end of the chain. These structures are important to health as they are essential nutrients used to make hormone-like substances that play regulatory roles in the body. (4) Saturated fats have all their carbon atoms loaded with hydrogen atoms; unsaturated fats have hydrogens removed; monounsaturated fats have two hydrogens removed and one double bond; polyunsaturated fats have multiple double bonds and several hydrogens missing. Structurally, a triglyceride consists of a "backbone" of glycerol with three fatty acids attached. (5) Hydrogenation adds hydrogens to unsaturated fats to reduce the number of double bonds and make fats more saturated and resistant to oxidation. Hydrogenation can increase a product's shelf life and add a desirable texture. A *trans*-fatty acid is one in which the hydrogens next to the double bonds are on opposite sides of the carbon chain; *trans*-fatty acids are thought to increase heart disease risk. *Trans*-fatty acids are thought to be involved in heart disease and experts sometimes advise that intakes of these fatty acids be limited. (6) Phospholipids have a choline or other phosphorus-containing acid in place of one of the fatty acids enabling them to function as emulsifiers in the body. The phospholipid allows the fatty acids to dissolve in water. Cholesterol is a sterol, and its C, H, and O atoms are arranged in rings. Cholesterol in the body can serve as starting material for many important body compounds. (7) Phospholipids are important parts of cell membranes, they help lipids move back and forth across the cell membranes into the watery fluids on both sides, and they enable fat-soluble vitamins and hormones to pass easily in and out of cells. Cholesterol's vital roles are many including providing energy and serving as part of cell membranes, bile acids, sex hormones, adrenal hormones, vitamin D. (8) Bile emulsifies fats allowing the enzymes to gain access to the fat for digestion; products of lipid digestion are packaged with protein for transport. Cholesterol shuttles back and forth between the liver and the body cells in lipoproteins, and it also visits the intestinal tract in the form of bile. (9) Composed of triglycerides, cholesterol, phospholipids, and proteins, lipoproteins transport lipids in the body. Chylomicrons are the largest of the lipoproteins, formed in the intestinal wall following fat absorption, and they contain mostly triglycerides; VLDL are made in the liver and contain mostly triglycerides; LDL contain few triglycerides but are about half cholesterol; HDL are about half protein and transport cholesterol back to the liver. (10) Linolenic acid and linoleic acid; vegetable oils and meats, grains, seeds, nuts, leafy vegetables, fish. (11) Excessive fat intake can contribute to elevated blood cholesterol and other blood lipids (therefore heart disease), obesity, cancer. Some saturated fats raise total cholesterol and LDL; polyunsaturated fatty acids lower LDL and HDL; *trans*-fatty acids raise LDL and lower HDL. (12) To consume a diet that is low in saturated fat, *trans* fat, and cholesterol (300 mg), limit total fat intake to 20 to 35% of daily energy from fat; consume 5 to 10 percent of daily energy from linoleic acid and 0.6 to 1.2 percent from linolenic acid. To reduce fat intake select lean meats and nonfat milk, eat plenty of vegetables, fruits, and grains, use fats and oils sparingly, look for invisible fat, and read food labels. (13) 65 grams; Daily Value for a person consuming 2000 kcalories per day. For a person who consumes 2000 kcalories a day, this number represents the recommended number of grams of fat for a day.

Short Answers—
1. triglycerides; phospholipids; sterols
2. maintain cell structure; protect organs; provide energy; protect lean tissue from depletion
3. to provide palatability; satiety; fat-soluble vitamins; kcalories
4. glycerol; 3 fatty acids
5. chylomicron; VLDL; LDL; HDL

Chemical Structure Identification:
1. glycerol
2. stearic acid
3. a polyunsaturated fatty acid (linoleic acid)
4. linoleic acid (an omega-6 fatty acid)
5. linolenic acid (an omega-3 fatty acid)
6. cholesterol

Sample Test Questions—
1. e (p. 141)
2. a (p. 141)
3. e (p. 165, 166)
4. d (p. 145)
5. b (p. 143)
6. b (p. 156)
7. c (p. 146)
8. c (p. 145, 146)
9. b (p. 146)
10. b (p. 148)
11. d (p. 148)
12. b (p. 149)
13. c (p. 152, 153)
14. a (p. 153)
15. d (p. 154)
16. c (p. 159)
17. c (p. 174)
18. d (p. 149)
19. a (p. 163)

Crossword Puzzle—

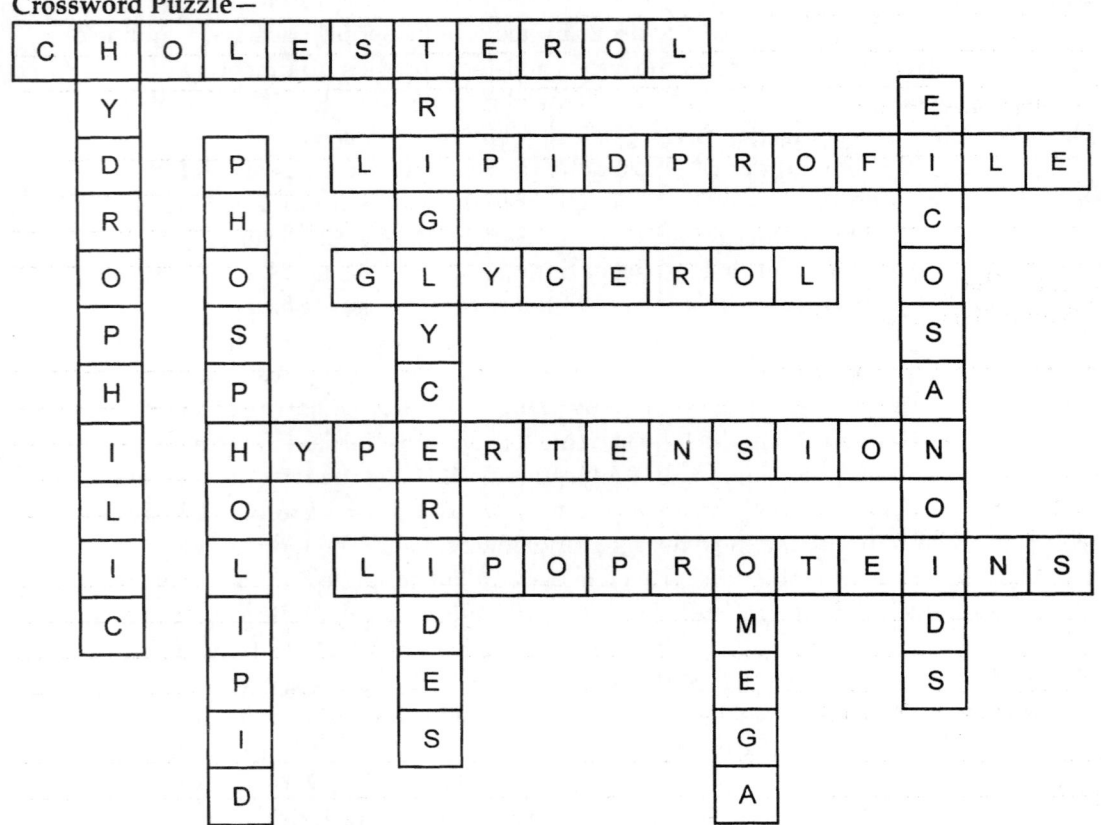

C	H	O	L	E	S	T	E	R	O	L			
Y					R							E	
D		P		L	I	P	I	D	P	R	O	F	I	L	E
R		H			G							C	
O		O		G	L	Y	C	E	R	O	L	O	
P		S			Y							S	
H		P			C							A	
I		H	Y	P	E	R	T	E	N	S	I	O	N
L		O			R							O	
I		L		L	I	P	O	P	R	O	T	E	I	N	S
C		I			D				M			D	
		P			E				E			S	
		I			S				G				
		D							A				

70

Chapter 6
Protein: Amino Acids

Chapter Outline

I. The Chemist's View of Proteins
    A. Amino Acids

    B. Proteins

II. Digestion and Absorption of Protein
    A. Protein Digestion

    B. Protein Absorption

III. Proteins in the Body
    A. Protein Synthesis

    B. Roles of Proteins

    C. A Preview of Protein Metabolism

IV. Protein in Foods
   A. Protein Quality

   B. Protein Regulations for Food Labels

V. Health Effects and Recommended Intakes of Protein
   A. Protein-Energy Malnutrition

   B. Health Effects of Protein

   C. Recommended Intakes of Protein

   D. Protein and Amino Acid Supplements

Highlight: Vegetarian Diets

SUMMING UP

Proteins are composed of (1)_____ acids. Nonessential amino acids can be (2)_____ in the body from carbohydrate derivatives and an amino nitrogen source. (3)_____ amino acids cannot be synthesized in the body, or cannot be made in amounts sufficient to meet physiological need. The major role of dietary protein is to supply amino acids for the synthesis of (4)_____ needed in the body, although dietary protein can also serve as an (5)_____ source.

A (6)_____ protein supplies all the essential amino acids; a high-quality protein not only supplies them, but also provides them in the appropriate (7)_____. (8)_____ protein sources are generally of higher quality than vegetable protein sources, but diets composed of plant foods provide plenty of protein, as long as a variety of nutrient-dense foods supply a sufficient energy intake.

Proteins act as (9)_____ and they help regulate the water balance and the (10)_____-_____ balance. (11)_____ and some hormones are made of proteins.

The absorption of many nutrients from the GI tract depends on protein (12)_____. The body's oxygen carrier, (13)_____ and the muscle's oxygen reservoir, myoglobin, are also proteins. Other body proteins include blood (14)_____ factors, collagen, and the light-sensitive (15)_____ of the retina. Maintenance and repair of body tissue and growth of new tissue requires a continual supply of (16)_____ acids to synthesize proteins.

Kwashiorkor is a disease in which (17)_____ is lacking. Protein deficiency combined with inadequate food (18)_____ is marasmus. These deficiencies are called protein-energy (19)_____ (PEM) and are a worldwide malnutrition problem.

In the exchange system, the foods that supply protein in abundance are on the (20)_____ list and the meat list. Vegetable and (21)_____ exchanges contribute some protein. Most people's average consumption of protein is considerably (22)_____ than the RDA. Diets especially high in protein may actually be (23)_____.

Assignments

Answer these chapter study questions from the text:

1. How does the chemical structure of proteins differ from the structures of carbohydrates and fats?

2. Describe the structure of amino acids, and explain how their sequence in proteins affects the proteins' shapes. What are the essential amino acids?

3. Describe protein digestion and absorption.

4. Describe protein synthesis.

5. Describe some of the roles proteins play in the human body.

6. What are enzymes? What roles do they play in chemical reactions? Describe the differences between enzymes and hormones.

7. How does the body use amino acids? What is deamination? Define nitrogen balance. What conditions are associated with zero, positive, and negative balance?

8. What factors affect the quality of dietary protein? What is a complete protein?

9. How can vegetarians meet their protein needs without eating meat?

10. What are the health consequences of ingesting inadequate protein and energy? Describe marasmus and kwashiorkor. How can the two conditions be distinguished, and in what ways do they overlap?

11. How might protein excess, or the type of protein eaten, influence health?

12. What factors are considered in establishing recommended protein intakes?

13. What are the benefits and risks of taking protein and amino acid supplements?

CHAPTER GLOSSARY

acidosis: above-normal acidity in the blood and body fluids.
acids: compounds that release hydrogen ions in a solution.
acute PEM: protein-energy malnutrition caused by recent severe food restriction; characterized in children by thinness for height (wasting).
aflatoxin: potent cancer-causing toxin produced by the mold *Aspergillus flavus* that infects grains and peanuts.
alkalosis: above-normal alkalinity (base) in the blood and body fluids.
amino acid pool: the supply of amino acids derived from either food proteins or body proteins that collect in the cells and circulating blood and stand ready to be incorporated in proteins and other compounds or used for energy.
amino acid scoring: a method of evaluating protein quality by comparing a test protein's amino acid pattern with that of a reference protein, sometimes called *chemical scoring*.
amino acids: building blocks of proteins; each contains an amino group, an acid group, a hydrogen atom, and a distinctive side group attached to a central carbon atom.
antibodies: large proteins of the blood and body fluids, produced by the immune system in response to the invasion of the body by foreign molecules.
antigens: substances that elicit the formation of antibodies or an inflammation reaction from the immune system.
bases: compounds that accept hydrogen ions in a solution.
biological value (BV): the amount of protein nitrogen that is retained for growth and maintenance, expressed as a percentage of the protein nitrogen that has been digested and absorbed; a measure of protein quality.
buffers: compounds that help keep a solution's acidity or alkalinity constant.
chronic PEM: protein-energy malnutrition caused by long-term food deprivation; characterized in children by short height for age (stunting).
collagen: the protein material from which connective tissues such as scars, tendons, ligaments, and the foundations of bones and teeth are made.
complementary proteins: two or more proteins whose amino acid assortments complement each other in such a way that the essential amino acids missing from one are supplied by the other.
complete protein: a protein containing all the amino acids essential in human nutrition in amounts adequate for human use.
conditionally essential amino acid: an amino acid that is normally nonessential, but must be supplied by the diet in special circumstances when the need for it exceeds the body's ability to produce it.
deamination: removal of the amino (NH_2) group from a compound such as an amino acid.
denaturation: the change in a protein's shape brought about by heat, acid, base, alcohol, heavy metals, or other agents.

digestibility: a measure of the amount of amino acids absorbed from a given protein intake.
dipeptide: two amino acids bonded together.
dysentery: an infection of the digestive tract that causes diarrhea.
edema: the swelling of body tissue caused by excessive amounts of fluids in the interstitial spaces; seen in protein deficiency.
endogenous protein: the protein in the body.
enzymes: proteins that facilitate chemical reactions without being changed in the process; protein catalysts.
essential amino acids: amino acids that the body cannot synthesize in amounts sufficient to meet physiological needs; sometimes referred to as indispensable.
exogenous protein: protein in foods.
ferritin: the protein residing in the cells of the intestinal wall; the storage protein.
fluid balance: maintenance of the proper types and amounts of fluid and minerals in each compartment of the body fluids.
hemoglobin: a globular protein in red blood cells that carries oxygen from the lungs to the cells throughout the body.
high-quality protein: an easily digestible, complete protein.
hormones: chemical messengers. Hormones are secreted by a variety of endocrine glands in response to altered conditions in the body. Each travels to one or more specific target tissues or organs, where it elicits a specific response.
immunity: the body's ability to recognize and eliminate foreign invaders.
kwashiorkor: a form of PEM that results from either inadequate protein intake, or more commonly, from infections.
limiting amino acid: the essential amino acid found in the shortest supply relative to the amounts needed for protein synthesis in the body.
marasmus: a form of PEM that results from a severe deprivation, or impaired absorption, of energy, protein, vitamins, and minerals.
matrix: the basic substance that gives form to a developing structure; in the body, the formative cells from which teeth and bones grow.
mutual supplementation: the strategy of combining two protein foods in a meal so that each food provides the essential amino acid(s) lacking in the other.
myoglobin: the muscle cell protein.
negative nitrogen balance: N in < N out.
net protein utilization (NPU): the amount of protein nitrogen that is retained from a given amount of protein nitrogen eaten; a measure of protein quality.
neurotransmitters: chemicals that are released at the end of a nerve cell when a nerve impulse arrives there; they diffuse across the gap to the next cell and alter the membrane of that second cell to either inhibit or excite it.
nitrogen balance: the amount of nitrogen consumed as compared with the amount of nitrogen excreted in a given period of time.
nitrogen equilibrium: zero nitrogen balance; N in = N out.
nonessential amino acids: amino acids that the body can synthesize; sometimes referred to as dispensable.
oligopeptide: an intermediate string of four to nine amino acids bonded together.
pepsin: a gastric protease.
peptidase: a digestive enzyme that hydrolyzes peptide bonds.
peptide bond: a bond that connects one amino acid with another, forming a link in a protein chain.
polypeptide: many (ten or more) amino acids bonded together.

positive nitrogen balance: N in > N out.
proteases: enzymes that hydrolyze protein.
protein digestibility: a measure of the amount of amino acids absorbed from a given protein intake.
protein efficiency ratio (PER): a measure of protein quality assessed by determining how well a given protein supports weight gain in growing rats; used to establish the protein quality for infant formulas and baby foods.
protein turnover: the degradation and synthesis of endogenous protein.
protein-digestibility-corrected amino acid score (PDCAAS): a measure of protein quality assessed by comparing the amino acid balance of a food protein with the amino acid requirements of preschool-aged children and then correcting for the true digestibility of protein.
protein-energy malnutrition (PEM), also called **protein-kcalorie malnutrition (PCM):** a deficiency of both protein and energy; the world's most widespread malnutrition problem, including kwashiorkor, marasmus, and instances in which they overlap.
proteins: compounds composed of carbon, hydrogen, oxygen, and nitrogen atoms, arranged into amino acids linked in a chain. Some amino acids also contain sulfur atoms.
reference protein: standard against which to measure the quality of other proteins.
sickle-cell anemia: a hereditary form of anemia characterized by abnormal sickle- or crescent-shaped red blood cells.
synthetase: an enzyme that enables two or more substances to form a more complex structure.
tripeptide: three amino acids bonded together.

Complete these short answer questions:

1. The atoms needed to make protein are:

    a.   c.

    b.   d.

2. The three common parts of all amino acids are:

    a.

    b.

    c.

3. Functions of proteins include:

    a.   f.

    b.   g.

    c.   h.

    d.   i.

    e.

4. The quality of dietary protein depends on:

   a.

   b.

5. The classic protein (-energy) deficiency diseases are:

   a.

   b.

Identify these chemical structures:

1.

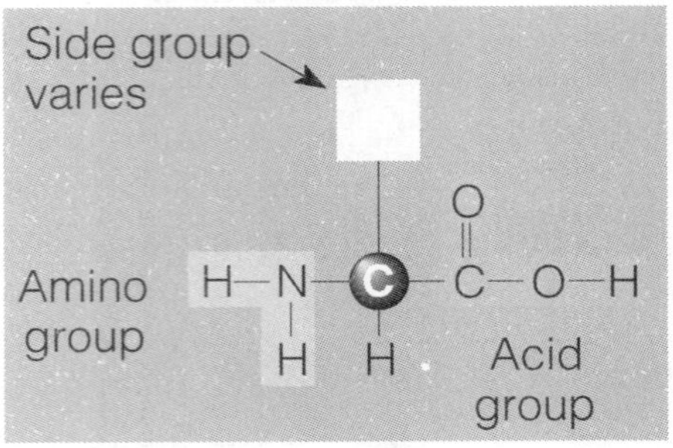

2.

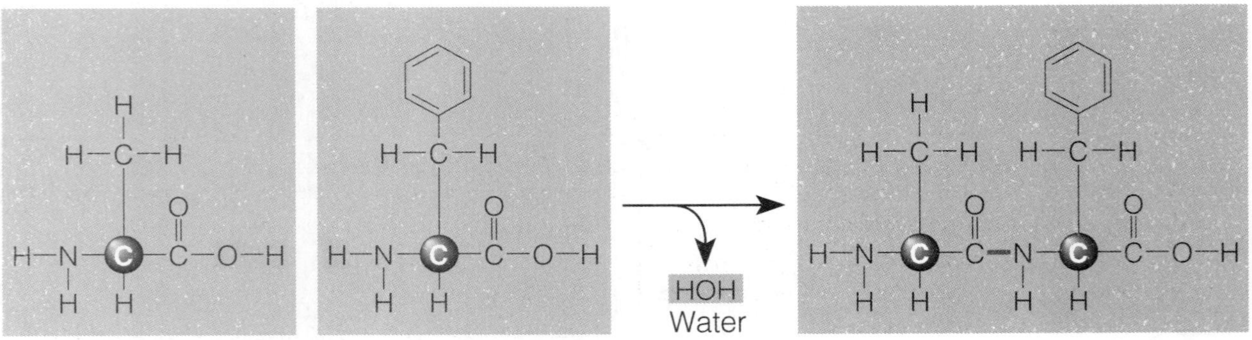

3.

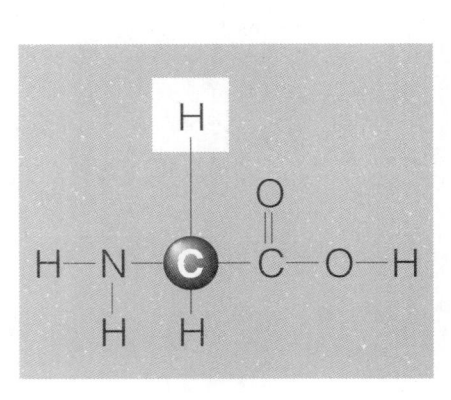

4.

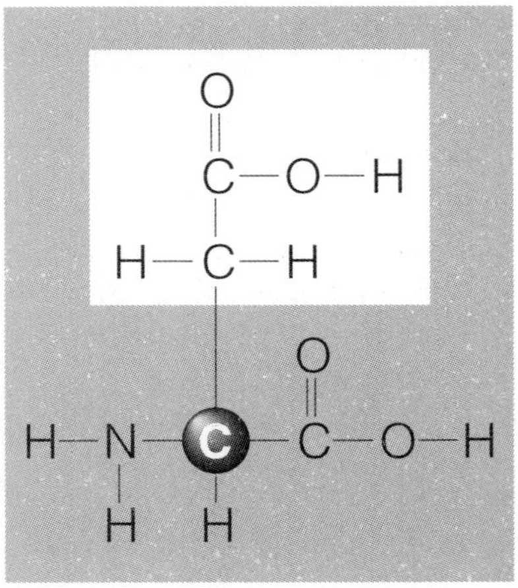

Complete this crossword puzzle by Mary A. Wyandt, Ph.D., CHES.

Across:	Down:
1. large proteins of the blood and body fluids, produced by the immune system in response to the invasion of the body by foreign molecules	2. chemical messengers secreted by a variety of endocrine glands in response to altered conditions in the body
4. the protein material from which connective tissues and the foundation of bones and teeth are made	3. the basic substance that gives form to a developing structure; in the body, the formative cells from which teeth and bones grow
8. proteins that facilitate chemical reactions without being changed in the process; protein catalysts	5. the swelling of body tissue caused by excessive amounts of fluids in the interstitial spaces
9. the body's ability to recognize and eliminate foreign invaders	6. compounds composed of carbon, hydrogen, oxygen, and nitrogen atoms, arranged into amino acids linked in a chain
10. a globular protein in red blood cells that carries oxygen from the lungs to the cells throughout the body	7. substances that elicit the formation of antibodies or an inflammation reaction form the immune system

Solve these problems:

1. Using the exchange system, how many grams of protein are in this meal?

   2 slices whole-wheat bread   1 tbsp mayonnaise
   1 oz boiled ham   1 c whole milk
   1 oz cheddar cheese   1 apple

2. How many kcalories does protein contribute to the above meal?

3. The meal provides 570 kcalories. What percentage of the kcalories is from protein?

4. What is the protein RDA for a person whose appropriate weight is 176 lb?

SAMPLE TEST QUESTIONS

1. Proteins differ from the other energy nutrients in that they contain:

   a. glycerol.   c. fatty acids.
   b. carbon.   d. nitrogen.

2. A polypeptide is:

   a. many amino acids linked together by peptide bonds.
   b. formation of a helix by adenine, thymine, guanine and cystosine.
   c. the primary structure of an amino acid.
   d. composed of glucose and amino acids.

3. The structures of amino acids differ in that:

   a. some do not contain nitrogen.
   b. some do not contain carbon.
   c. they each have a different side chain.
   d. they each have different acid groups.

4. When an enzyme is used in a chemical reaction in the body, it:

   a. forms ATP.
   b. remains unchanged.
   c. is degraded into a substance containing less chemical energy.
   d. only breaks down substances.

5. Proteins maintain acid-base balance in the body by:

   a. acting as buffers.
   b. secreting chloride ions.
   c. secreting acids.
   d. tying up excess sodium ions.

Match the following:

   _____ 6. enzyme   a. elicits the formation of antibodies.
   _____ 7. antibody   b. protein catalyst.
   _____ 8. hormone   c. inactivates foreign agents.
   _____ 9. opsin   d. a chemical messenger.
   _____ 10. antigen   e. the storage protein.
   _____ 11. transferrin   f. visual pigment protein.

12. Methionine, threonine, and tryptophan are names of:

    a. proteins.
    b. fatty acids.
    c. essential amino acids.
    d. lipids.
    e. nonessential amino acids.

13. If an essential amino acid required for formation of a certain enzyme is missing in the diet:

    a. another amino acid will be substituted in its place so the enzyme can be made.
    b. synthesis of the enzyme will stop.
    c. the partially synthesized enzyme will be stored in the adipose tissue until the missing amino acid is supplied in the diet.
    d. the amino acid will be made from glucose.

14. The new standard for the reference protein is:

    a. the most digestible proteins.
    b. amino acid scoring.
    c. meat protein.
    d. egg protein.
    e. the essential amino acid requirements of preschool-age children.

15. The amount of protein nitrogen that is retained from a given amount of protein nitrogen eaten is:

    a. nitrogen output.
    b. nitrogen intake.
    c. protein efficiency ration.
    d. net protein utilization

16. The most widespread form of malnutrition in developing countries is:

    a. anorexia.
    b. bulimia.
    c. protein-energy malnutrition.
    d. iron-deficiency anemia.
    e. vitamin C deficiency.

17. Marasmus may be described as a nutritional deficiency disease with:

    a. severe deficiencies of protein and vitamins.
    b. severe deficiency of protein.
    c. severe deficiencies of protein, energy, vitamins, and minerals.
    d. excessive amount of edema.

18. The Daily Value for protein based on a 2000 kcalorie diet is:

    a. 5 grams
    b. 10 grams
    c. 35 grams
    d. 50 grams

19. The food group which contributes zero protein is:

    a. milk.
    b. starchy vegetable.
    c. vegetable.
    d. fruit.

Answers

Summing Up—(1) amino; (2) synthesized; (3) Essential; (4) proteins; (5) energy; (6) complete; (7) proportions; (8) Animal; (9) enzymes; (10) acid-base; (11) Antibodies; (12) pumps; (13) hemoglobin; (14) clotting; (15) pigments; (16) amino; (17) protein; (18) energy; (19) malnutrition; (20) milk; (21) bread; (22) higher; (23) hazardous.

Chapter study questions from the text—(1) Like carbohydrates and fats, proteins contain carbon, hydrogen, and oxygen, but proteins also contain nitrogen. Amino acid structures have an amino group, an acid group, a hydrogen, and a side chain which makes each amino acid different from the others. (2) Amino acids are linked together to form proteins. The sequence and special characteristics of the side chains determine the protein's shape. The essential amino acids are histidine, isoleucine, leucine, lysine, methionine, phenylalanine, threonine, tryptophan, valine. (3) In the mouth, chewing and crushing moisten protein-rich foods and mix them with saliva to be swallowed; in the stomach, stomach acid uncoils protein strands and activates enzymes, pepsin and HCl break protein down into smaller polypeptides; in the small intestine, pancreatic and small intestinal enzymes split polypeptides further into dipeptides, tripeptides, and amino acids, then enzymes on the surface of the small intestinal cells hydrolyze these peptides and the cells absorb them. (4) DNA is in the nucleus of each cell; DNA serves as a template to make strands of messenger RNA. Each messenger RNA strand carries instructions for some protein the cell needs; the messenger RNA leaves the nucleus, and attaches itself to the protein-making machinery of the cell. Transfer RNA carry amino acids to the messenger RNA which dictates the sequence in which they will snap into place; this lines up the amino acids in sequence. The amino acids are then linked together in sequence, the completed protein strand is released, and later, the messenger

RNA is degraded and the transfer RNA are re-used. (5) Proteins serve as enzymes, help maintain the body's fluid balance by attracting water, help maintain acid-base balance by acting as buffers, act against disease agents as antibodies, regulate body processes as hormones, transport nutrients and other molecules into and out of cells, help clot blood, help make scar tissue and bones, and serve as light-sensitive visual pigments. (6) Protein catalysts that facilitate the synthesis of larger compounds from smaller ones and hydrolysis of larger compounds to smaller ones without being affected in the process. Hormones are chemical messengers that are secreted by a variety of endocrine glands in response to altered conditions in the body. (7) The body uses amino acids for proteins or nonessential amino acids, or for other compounds such as for synthesis of the neurotransmitters norepinephrine and epinephrine, or melanin, or energy. Deamination is removal of the amino group from a compound such as an amino acid. Nitrogen balance: the amount of nitrogen consumed compared with the amount of nitrogen excreted. Zero N balance: normal, healthy adult and lactating mothers; positive N balance: growing children and pregnant women; negative N balance: people who are sick or in trauma and people with kidney disease. (8) Its supply of a balance of the essential amino acids and its digestibility. A complete protein is a protein containing all the amino acids essential in human nutrition in amounts adequate for human use.
(9) Vegetarians can obtain protein from legumes, nuts, vegetables, grains and (in some cases) eggs and milk products. (10) Protein-energy malnutrition, poor growth in children and weight loss and wasting in adults. Maramus is the disease of starvation and kwashiorkor is the deficiency disease caused by inadequate protein in the presence of adequate food energy. The distinction is that kwashiorkor has adequate energy, while maramus has both inadequate protein and energy. Both conditions cause loss of body protein tissue. (11) Diets too high in protein offer no benefits. They often contribute fat-rich foods to diets of people who need to lose weight. In the small infant, the accumulation of amino acids stresses the kidneys and liver which have to metabolize and excrete the excess nitrogen; this may cause acidosis, dehydration, diarrhea, elevated blood ammonia, elevated blood urea, and fever. High-protein diets may promote calcium losses and deplete the bones of this mineral; if much animal-derived protein is eaten, this may contribute to the development of heart disease and cancer. (12) The quality of the protein eaten, kcalories and other nutrients consumed, a person's lean body mass, and a person's state of health. (13) No benefits, but risks include death, abnormal heart rhythms, severe amino acid imbalances and toxicities; taking tryptophan supplements can cause eosinophilia myalgia syndrome (EMS).

Short Answers—
1. carbon; hydrogen; oxygen; nitrogen
2. amino group (NH2); acid group (COOH); hydrogen (H)
3. enzymes; fluid balance; acid-base balance; antibodies; hormones; transport proteins; blood clotting; connective tissue; visual pigments
4. a balance of essential amino acids; protein digestibility
5. kwashiorkor; marasmus

Chemical Structure Identification—
1. amino acid structure
2. condensation of 2 amino acids to form a dipeptide
3. glycine
4. aspartic acid

Crossword Puzzle—

```
          H
A N T I B O D I E S
          R
          M
      C O L L A G E N         P
        N     M   D           R
        E     A   E           O
A             T               
E N Z Y M E S R   I M M U N I T Y
T             I               E
I             X   A           
G                             I
E         H E M O G L O B I N S
N                             S
S
```

Problem Solving—

1. 2 (3 g) + 7 g + 7 g + 0 g + 8 g + 0 g = 28 g
2. 4 kcal/g X 28 g = 112 kcal
3. 112 kcal divided by 570 kcal = 20%
4. 176 lb divided by 2.2 lb/kg = 80 kg; 0.8 g/kg X 80 kg = 64 g

Sample Test Questions--

1. d (p. 181)
2. a (p. 183)
3. c (p. 182)
4. b (p. 190)
5. a (p. 191, 192)
6. b (p. 190)
7. c (p. 192)
8. d (p. 190)
9. f (p. 192)
10. a (p. 192)
11. e (p. 191)
12. c (p. 195)
13. b (p. 195)
14. e (p. 195)
15. d (p. 183)
16. c (p. 197)
17. c (p. 198)
18. d (p. 197)
19. d (p. 196)

Chapter 7
Metabolism: Transformations and Interactions

Chapter Outline

I. Chemical Reactions in the Body

II. Breaking Down Nutrients for Energy
    A. Glucose

    B. Glycerol and Fatty Acids

    C. Amino Acids

    D. Breaking Down Nutrients for Energy—In Summary

    E. The Final Steps of Catabolism

III. The Body's Energy Budget
    A. The Economics of Feasting

    B. The Transition from Feasting to Fasting

C. The Economics of Fasting

_____
_____
_____
_____

Highlight: Alcohol and Nutrition
_____
_____
_____
_____
_____

SUMMING UP

The principal compounds derived from carbohydrate, (1)_____, and protein in the diet are (2)_____, glycerol, fatty acids, and amino acids. Glucose may be anabolized to (3)_____ or catabolized to (4)_____, which in turn yields acetyl CoA. Glycerol and fatty acids may be anabolized to (5)_____ or catabolized—glycerol to pyruvate, fatty acids to acetyl CoA. Amino acids may be anabolized to (6)_____ or catabolized (after deamination) to pyruvate, acetyl CoA, or intermediates that enter the (7)_____ cycle directly. When amino acids are deaminated, the nitrogen removed is combined with carbon dioxide to form urea and excreted.

Pyruvate is convertible to glucose, but (8)_____ CoA is not. Hence fatty acids, which break down to multiple units of acetyl CoA, cannot serve to generate (9)_____ for the body. All three energy-yielding nutrients are convertible to acetyl CoA, however, and hence can be used to manufacture body (10)_____, or provide (11)_____.

If a person (12)_____ or if carbohydrate is undersupplied, lean body tissue is catabolized to meet the brain's need for (13)_____. Within a day or so after the start of a fast, a person is in (14)_____. (15)_____ acids are metabolized to ketone bodies, which can meet some of the (16)_____ energy need. Weight loss may be dramatic, but (17)_____ loss may be slower than on a moderate, balanced low-kcalorie diet. To design such a moderate diet requires adjusting energy balance so that the intake is (18)_____, the output is (19)_____, or both.

ASSIGNMENTS

Answer these chapter study questions from the text:

1. Define metabolism, anabolism, and catabolism; give an example of each.

2. Name one of the body's high-energy molecules, and describe how is it used.

3. What are coenzymes, and what service do they provide in metabolism?

4. Name the four basic units, derived from foods, used by the body in metabolic transformations. How many carbons are in the "backbones" of each?

5. Define aerobic and anaerobic metabolism. How does insufficient oxygen influence metabolism?

6. How does the body dispose of excess nitrogen?

7. Summarize the main steps in the metabolism of glucose, glycerol, fatty acids, and amino acids.

8. Describe how a surplus of the three energy nutrients contributes to body fat stores.

9. What adaptations does the body make during a fast? What are ketone bodies? Define ketosis.

10. Distinguish between a loss of *fat* and a loss of *weight*, and describe how both might happen.

Chapter Glossary

acetyl CoA: a 2-carbon compound (acetate, or acetic acid) to which a molecule of CoA is attached.
aerobic: requiring oxygen.
ammonia: a compound with the chemical formula NH_3; produced during deamination.
anabolism: reactions in which small molecules are put together to build larger ones. Anabolic reactions require energy.
anaerobic: not requiring oxygen.
ATP (adenosine triphosphate): a common high-energy compound composed of a purine (adenine), a sugar (ribose), and three phosphate groups.
catabolism: reactions in which large molecules are broken down to smaller ones. Catabolic reactions usually release energy.
CoA: coenzyme A; the coenzyme derived from the B vitamin pantothenic acid and central to the energy metabolism of nutrients.
coenzymes: small organic molecules that work with enzymes to facilitate the enzymes' activity. Many coenzymes have B vitamins as part of their structures.
cofactors: substances that facilitate enzyme action; this includes both organic coenzymes such as vitamins and inorganic substances such as minerals.
Cori cycle: the path from muscle glycogen to glucose to pyruvate to lactic acid (which travels to the liver) to glucose (which can travel back to the muscle) to glycogen, named after the scientist who elucidated this pathway.
coupled reactions: pairs of chemical reactions in which energy released from the breakdown of one compound is used to create a bond in the formation of another compound.
electron transport chain: the final pathway in energy metabolism where the electrons from hydrogen are passed to oxygen and the energy released is trapped in the bonds of ATP.
energy metabolism: the chemical reactions by which the body obtains and spends energy.
fatty acid oxidation: the metabolic breakdown of fatty acids to acetyl CoA.
fuel: compounds that cells can use for energy.
glycolysis: the metabolic breakdown of glucose to pyruvate. Glycolysis does not require oxygen.
keto acid: an organic acid that contains a carbonyl group (C=O).
lactic acid: an acid produced from pyruvate during anaerobic metabolism.
metabolism: the sum total of all the chemical reactions that go on in living cells.
mitochondria: the cellular organelles responsible for producing ATP aerobically.
oxaloacetate: a carbohydrate intermediate of the TCA cycle.
photosynthesis: the process by which green plants make carbohydrates from carbon dioxide and water using the green pigment chlorophyll to trap the sun's energy.
pyruvate: pyruvic acid, a 3-carbon compound that, in metabolism, can be derived from glucose, certain amino acids, or glycerol.
TCA cycle: tricarboxylic acid cycle; a series of metabolic reactions that break down molecules of acetyl CoA to carbon dioxide and hydrogen atoms.
transamination: the transfer of an amino group from one amino acid to a keto acid, producing a new nonessential amino acid and a new keto acid.
urea: the principal nitrogen-excretion product of metabolism. Two ammonia fragments are combined with carbon dioxide to form urea.

Complete these short answer questions:

1. The energy nutrients are broken down into these basic units:

    a. carbohydrate: _____   c. lipids: _____

    b. lipids: _____   d. protein: _____

2. How many carbons are in these compounds?

    a. glucose: _____   d. amino acids: _____

    b. glycerol: _____   e. pyruvate: _____

    c. fatty acids: _____   f. acetyl CoA: _____

Identify the following structures, pathways, or processes:

1.

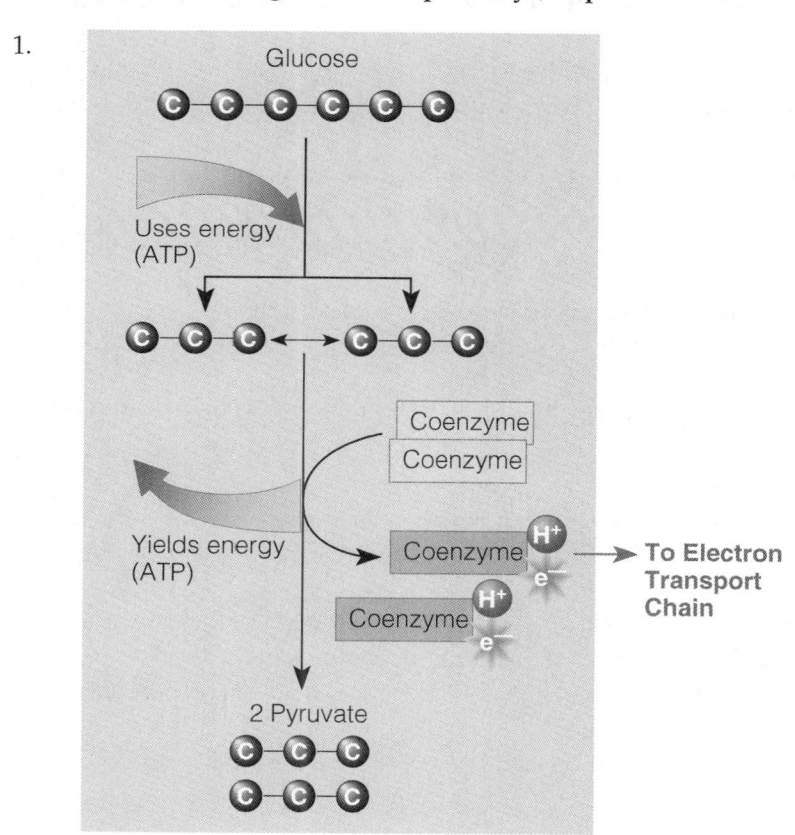

2.

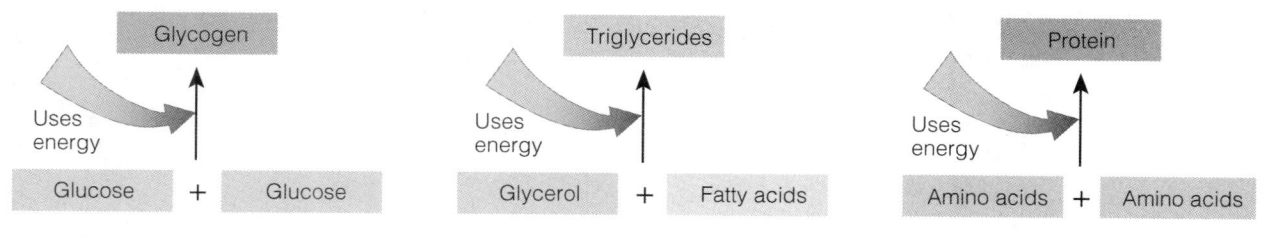

3.

4.

Solve these problems:

1. To lose 3 pounds, a person needs to adjust the energy budget so that intake is less than output by how many kcalories?

2. What should the daily kcalorie deficit be if a person wanted to lose 3 pounds in 30 days?

3. If the person were to consume 750 kcalories less per day than the energy output, how long would it take to lose 3 pounds?

Complete this crossword puzzle by Mary A. Wyandt, Ph.D., CHES.

Across:	Down:
1. substances that facilitate enzyme action; this includes both organic coenzymes and inorganic substances 3. a series of metabolic reactions that break down molecules of acetyl CoA to CO₂ and hydrogen atoms 4. an acid produced from pyruvate during anaerobic metabolism 8. the process by which green plants make carbohydrates from carbon dioxide and water using the green pigment chlorophyll to trap the sun's energy 9. an organic acid that contains carbonyl group (C=O)	1. pairs of chemical reactions in which energy released from the breakdown of one compound is used to create a bond in the formulation of another 2. a carbohydrate intermediate of the TCA cycle 5. produced during deamination, a compound with the chemical formula NH₃ 6. the path from muscle glycogen to glucose to pyruvate to lactic acid 7. pyruvic acid, a 3-carbon compound, that in metabolism, can be derived form glucose, certain amino acids, or glycerol

SAMPLE TEST QUESTIONS

1. Nutrients which are oxidized in the body to yield energy are:

    a. water, carbohydrate, fat, protein, vitamins, and minerals.
    b. carbohydrate, fat, protein, vitamins, and minerals.
    c. carbohydrate, fat, protein, and vitamins.
    d. carbohydrate, fat, and protein.
    e. carbohydrates and vitamins.

2. Reactions in which compounds are broken down into simple molecules are called ____ reactions.

    a. anabolic
    b. catabolic
    c. gluconeogenic
    d. erogenic

3. Small organic molecules that work to facilitate activity of enzymes are:

    a. adenosines.
    b. phosphates.
    c. coenzymes.
    d. acetyl CoA.

4. A three-carbon compound reversibly convertible to glucose is:

    a. amino acid.
    b. pyruvate.
    c. acetyl CoA.
    d. fatty acid.
    e. glucose.

5. When glucose is split in half, it yields two molecules of:

    a. acetate.
    b. glycerol.
    c. pyruvate.
    d. water.
    e. monosaccharides.

6. The part(s) of triglycerides that can be made into glucose is(are):

    a. short-chain fatty acids.
    b. medium-chain fatty acids.
    c. long-chain fatty acids.
    d. all fatty acids.
    e. glycerol.

7. The reaction which removes the NH_2 group from an amino acid is called:

    a. transamination.
    b. deamination.
    c. nitrogenation.
    d. hydrogenation.

8. Nitrogen from amino groups is excreted from the body in:

    a. urea.
    b. ammonia.
    c. nitrogen gas.
    d. amino acids.

9. When glucose consumption is in excess of body needs, the excess glucose is:

    a. not absorbed from the small intestine.
    b. excreted in the feces.
    c. stored as glucose.
    d. stored as glycogen only.
    e. stored as glycogen and fat.

10. Energy is stored in the body for future use as:

    a. triglycerides.
    b. glycerol.
    c. cholesterol.
    d. lecithin.

11. Unfortunately, food energy surpluses cannot be stored in the body.

    a. true
    b. false

12. Glucose is the preferred fuel for cells of the brain and nervous system.

    a. true
    b. false

13. During the second phase of fasting the body adapts by producing an alternative energy source:

    a. glycogen.
    b. ammonia.
    c. pyruvate.
    d. ketone bodies.

14. Symptoms of starvation include all but one of the following:

    a. wasting.
    b. raised metabolism.
    c. lowered body temperature.
    d. reduced resistance to disease.

15. Prolonged fasting results in the slowing of metabolism and all but one of the following is observed:

    a. muscle wasting.
    b. significant loss of body fat.
    c. significant loss of water.
    d. significant loss of body weight.

ANSWERS

Summing Up—(1) fat; (2) glucose; (3) glycogen; (4) pyruvate; (5) triglycerides; (6) protein; (7) TCA; (8) acetyl; (9) glucose; (10) fat; (11) energy; (12) fasts; (13) glucose; (14) ketosis; (15) Fatty; (16) brain's; (17) fat; (18) reduced; (19) increased.

Chapter study questions from the text—(1) Metabolism includes all chemical reactions, both anabolic and catabolic, that occur in living cells; for example, fructose is metabolized to glucose. Anabolic reactions put small molecules together to build larger ones; for example, glycerol + fatty acids = triglyceride. Catabolic reactions break down large molecules to smaller ones; for example, triglyceride = glycerol + fatty acids. (2) ATP is used as energy currency in metabolic reactions. (3) Coenzymes are small organic molecules that work with enzymes to facilitate the enzymes' activity; they enable enzymes to do their work. (4) Glucose, glycerol, fatty acids, and amino acids. Glucose has 6 carbons, glycerol has 3 carbons, fatty acids have multiples of 2 carbons, and amino acids have 2, 3, or more carbons. (5) Aerobic metabolism refers to chemical reactions that require oxygen; anaerobic reactions do not require oxygen. Insufficient oxygen favors anaerobic metabolism that results in lactic acid production. (6) When amino nitrogen is stripped from amino acids, ammonia is produced. The liver detoxifies ammonia before releasing it into the bloodstream by combining it with another waste product, carbon dioxide, to produce urea. (7) Glucose to pyruvate to acetyl CoA to carbon dioxide. Glycerol to pyruvate either to glucose or to acetyl CoA to carbon dioxide. Fatty acids either to pyruvate or to acetyl CoA to carbon dioxide. (8) Each can be broken down to pyruvate and acetyl CoA which can be built up into fat (fat can also be stored directly). (9) The body draws on carbohydrate and fat reserves to spare protein; when reserves are exhausted, body protein is used to make glucose; as fast continues, ketones are made to meet energy needs; body reduces its energy output and fat loss. Ketone bodies are compounds produced during the incomplete breakdown of fat when glucose is not available. Ketosis is the buildup of ketone bodies in the blood and urine. (10) Weight loss on a low-carbohydrate diet is a loss of glycogen, protein, water, and minerals, not just a loss of fat. A moderate, balanced low-kilocalorie diet is more likely to promote loss of fat.

Short Answers—
1. glucose; glycerol; fatty acids; amino acids
2. 6; 3; multiples of 2; 2 or 3 or more; 3; 2

Identification of structures, pathways, or processes:
1. glycolysis
2. anabolic reactions
3. ATP
4. urea synthesis

Problem Solving--
1. 3 lb X 3,500 kcal/lb = 10,500 kcal
2. 10,500 kcal divided by 30 days = 350 kcal/day
3. 10,500 kcal divided by 750 kcal/day = 14 days

Crossword Puzzle—

Across: COFACTORS, TCACYCLE, LACTICACID, PHOTOSYNTHESIS, KETOACID

Down: COUPLEDREACTIONS, OXALOACETATE, CITRICACYCLE, AMMONIA, PYRUVATE

Sample Test Questions—

1. d (p. 216)
2. b (p. 216)
3. c (p. 219)
4. b (p. 221)
5. c (p. 221)
6. e (p. 224)
7. b (p. 226, 227)
8. a (p. 228)
9. e (p. 232)
10. a (p. 232)
11. b (p. 228)
12. a (p. 236)
13. d (p. 236)
14. b (p. 237)
15. b (p. 237)

CHAPTER 8
ENERGY BALANCE AND BODY COMPOSITION

CHAPTER OUTLINE

I. Energy Balance

II. Energy In: The kCalories Foods Provide
    A. Food Composition

    B. Food Intake

III. Energy Out: The kCalories the Body Spends
    A. Components of Energy Expenditure

    B. Estimating Energy Requirements

IV. Body Weight, Body Composition, and Health
    A. Defining Healthy Body Weight

    B. Body Fat and Its Distribution

    C. Health Risks Associated With Body Weight and Body Fat

Highlight: The Latest and Greatest Weight-Loss Diet—Again

_____
_____
_____
_____

SUMMING UP

Energy intake can be computed by adding up the (1)_____ values of the foods consumed or can be estimated by using (2)_____ system values. The energy available from food was originally determined by measuring the (3)_____ lost when the food was completely burned, with adjustments to account for the incomplete breakdown of food in the (4)_____. Machines designed to measure energy expenditure work by measuring heat lost from the body or (5)_____ and carbon dioxide exchanged. Data collected in this way have yielded tables from which to estimate (6)_____ needs.

The body's total energy output falls into two major categories: energy to support the (7)_____ or resting metabolism, and energy for (8)_____ activity. The basal metabolic rate is influenced primarily by the amount of (9)_____ body mass; it is also affected by fever, fasting, and (10)_____ secretions (especially epinephrine and thyroxin).

An energy deficit of (11)_____ kcalories is necessary for the loss of a pound of body fat. Loss of body fat in excess of (12)_____ pounds a week can rarely be sustained. A diet supplying fewer than 1,200 kcalories per day can be made adequate in (13)_____ and minerals only with great difficulty. People of the same sex, age, and height may differ in (14)_____ not only due to differing densities of their bones and muscles. The weight compatible with good health depends on the individual. Obesity is sometimes defined as body weight more than (15)_____ percent above desirable weight. More precisely, obesity is (16)_____ body fatness.

Obesity entails a host of health hazards, in addition to (17)_____ and economic disadvantages. By contrast, (18)_____ (weight more than 10 percent below desirable weight) is associated with an increased risk from wasting diseases. It is likely that both (19)_____ and environmental factors influence obesity in human beings. Abnormalities in hunger or (20)_____ regulation may cause obesity; but probably the most important contributor to obesity in this country is the extreme physical (21)_____ that characterizes a sedentary lifestyle.

Chapter Glossary

adaptive thermogenesis: adjustments in energy expenditure related to changes in environment such as cold and to physiological events such as overfeeding, trauma, and changes in hormone status.
air displacement plethysmography: estimates body composition by having a person sit inside a chamber while computerized sensors determine the amount of air displaced by the person's body.
appetite: the psychological desire to eat or interest in food; a positive sensation that accompanies the sight, smell, or thought of food.
basal metabolic rate (BMR): the rate of energy use for metabolism under basal conditions, usually expressed as kcalories per kilogram body weight per hour.
basal metabolism: the energy needed to maintain life when a body is at complete rest after a 12-hour fast (to exclude the thermic effect of the previous meal).
bioelectrical impedance: a method for estimating body fat using low-intensity electrical current.
body composition: the proportions of muscle, bone, fat, and other tissue that make up a person's total body weight.
body mass index (BMI): an index of a person's weight in relation to height, determined by dividing the weight (in kilograms) by the square of the height (in meters).
bomb calorimeter: an instrument that measures the *heat* energy released when foods are burned, thus providing an estimate of the potential energy of foods.
central obesity: excess fat around the trunk of the body; also called *abdominal fat* or *upper body fat*.
direct calorimetry: the measurement of energy output as heat energy.
dual energy X-ray absorptiometry (DEXA): uses two low-dose X-rays that differentiate among fat-free soft tissue (lean body mass), fat tissue, and bone tissue, providing a precise measurement of total fat and its distribution in all but extremely obese people.
external cue theory: the theory that some people eat in response to such external factors as the presence of food or the time of day rather than to such internal factors as hunger.
fatfold measure: a clinical estimate of total body fatness in which the thickness of a fold of skin on the back of the arm (over the triceps muscle), below the shoulder blade (subscapular), or in other places is measured with a caliper; also called *skinfold test*.
frame size: the size of a person's bones and musculature.
hunger: the physiological need to eat, experienced as a drive to obtain food; an unpleasant sensation.
hydrodensitometry: a method of measuring body density in which the person is first weighed and then submerged in water.
hypothalamus: a brain center that controls activities such as maintenance of water balance, regulation of body temperature, and control of appetite.
indirect calorimetry: the estimation of energy output from measures of the amount of oxygen used and carbon dioxide eliminated.
insulin resistance: the reduced ability of insulin to regulate glucose metabolism.
intra-abdominal fat: fat stored within the abdominal cavity in association with the internal abdominal organs, as opposed to the fat stored directly under the skin (subcutaneous fat).
lean body mass: the weight of the body minus the fat content.
neuropeptide Y: a chemical produced in the brain that stimulates appetite, diminishes energy expenditure, and increases fat storage.
overfat: an excess of body fat.
overweight: body weight above some standard of acceptable weight that is usually defined in relation to height (such as the weight-for-height tables).
physiological fuel value: the number of kcalories that the human body derives from a food, as contrasted with the number of kcalories determined by calorimetry.

resting metabolic rate (RMR): a measure of a person at rest in a comfortable setting, but with less stringent criteria for the number of hours fasting.
satiation: the feeling of satisfaction that occurs during a meal and halts eating.
satiating: having the power to suppress hunger and inhibit eating.
satiety: the feeling of satisfaction and fullness that food brings.
stress eating: eating in response to arousal.
thermic effect of food (TEF): an estimation of the energy required to process food (digest, absorb, transport, metabolize, and store ingested nutrients); also called *diet induced thermogenesis (DIT)*, the *specific dynamic effect (SDE)* of food, or the *specific dynamic activity (SDA)* of food.
thermogenesis: the generation of heat; used in physiology and nutrition studies as an index of how much energy the body is spending.
underweight: body weight below some standard of acceptable weight that is usually defined in relation to height (such as the weight-for-height tables).
voluntary activities: the component of a person's daily energy expenditure that involves conscious and deliberate muscular work (walking, lifting, climbing, or other physical activity).
waist circumference: an anthropometric measurement used to assess a persons abdominal fat.

ASSIGNMENTS

Answer these chapter study questions from the text:

1. What are the consequences of an unbalanced energy budget?

2. Define hunger, appetite, satiation, and satiety and describe how each influences food intake.

3. Describe each component of energy expenditure. What factors influence each? How can energy expenditure be estimated?

4. Distinguish between body weight and body composition. What assessment techniques are used to measure each?

5. What problems are involved in defining "ideal" body weight?

6. What is central obesity and what is its relationship to disease?

7. What risks are associated with excess body weight and excess body fat?

Complete this short answer question:

1. The two major contributors to energy output are:

   a.

   b.

Solve this problem:

1. Calculate the energy output for basal metabolism for a 175-pound man.

Complete this crossword puzzle by Mary A. Wyandt, Ph.D., CHES.

Across:	Down:
1. a method of measuring body density; the person is first weighed and then submerged in water	2. a generation of heat; used in physiology and nutrition studies as an index of how much energy the body is spending
3. a clinical estimate of total body fatness in which the thickness of a fold of skin is measured with a caliper	4. a brain center that controls activities such as maintenance of water balance, regulation of body temperature, and control of appetite
7. a type of calorimetry measurement that estimates the energy output from measures of the amount of oxygen used and carbon dioxide eliminated	5. the weight of the body minus the fat content
	6. a type of calorimetry measurement that measures energy output as heat energy
9. the proportions of muscle, bone, fat, and other tissue that make up a person's total body weight	8. abbreviation for an index of a person's weight in relation to height, determined by dividing the weight (kg) by the square of the height (m)
10. the energy needed to maintain life when a body is at complete rest after a 12-hour fast	

Sample Test Questions

1. For each _____ kcal eaten in excess, one pound of body fat is stored.

   a. 2,000
   b. 2,500
   c. 3,000
   d. 3,500

2. The feeling of fullness that occurs during a meal and halts eating is called:

   a. satiation
   b. satiety
   c. purging
   d. appetite

3. Which of the following describes the process of thermogenesis?

   a. burning of fat
   b. synthesis of fat
   c. generation of heat
   d. generation of water

4. What method is used to measure the amount of heat given off by the body?

   a. bomb calorimetry
   b. basal calorimetry
   c. direct calorimetry
   d. indirect calorimetry

5. Additional energy that is spent when a person must adapt to extremely cold conditions is called:

   a. basal metabolic rate.
   b. resting metabolism.
   c. adaptive thermogenesis.
   d. rebound thermogenesis.

6. The amount of energy a food contains may be measured by burning the food and measuring the:

   a. vitamins produced.
   b. heat released.
   c. water produced.
   d. water released.
   e. weight of the ash remaining.

7. The energy used by the body completely at rest with no food recently eaten is:

   a. specific dynamic energy.
   b. basal metabolic energy.
   c. active energy.
   d. total kcalories expended.
   e. zero energy.

8. Which of the following will reduce BMR?

   a. fever
   b. fasting
   c. epinephrine
   d. high thyroid activity

9. A measure of energy output that is less precise than the BMR is:

   a. physical activity.
   b. thermic effect.
   c. specific dynamic effect.
   d. resting metabolic rate.

10. What is the major factor that determines metabolic rate?

    a. age
    b. gender
    c. amount of fat tissue
    d. amount of lean body mass

11. Which of the following increases the metabolic rate?

    a. fever
    b. fasting
    c. inactivity
    d. malnutrition

12. What would be the approximate weight gain of a person who consumes an excess of 500 kcal daily for one month?

    a. 0.5 lb.
    b. 2 lbs.
    c. 3 lbs.
    d. 4 lbs.

13. A BMI of 27 is considered:

    a. underweight.
    b. normal.
    c. overweight.
    d. obese.

14. If you eat 2,400 kcalories a day, about _____ will be used for the thermic effect of food.

    a. 24 kcalories
    b. 240 kcalories
    c. 100 kcalories
    d. 10 kcalories
    e. 3,500 kcalories

15. Degree of fatness can most accurately be assessed by:

    a. the fatfold test.
    b. accepted style of the culture.
    c. undressing and standing before a mirror to see how you look to yourself.
    d. the fit of your clothes.
    e. comparing weight with standard weight charts.

16. An index of a person's weight in relation to height is called:

    a. body mass index.
    b. height to weight index.
    c. ideal body weight index.
    d. desirable body weight index.

17. Which of the following is *not* a feature of the fatfold test?

    a. Can be self-administered.
    b. Correlates with risk of heart disease.
    c. Provides a direct estimate of the amount of body fat.
    d. Considered a practical diagnostic tool for trained users.

18. The method of body fat assessment that requires the person to be submerged in water is called:

    a. fat fold.
    b. hydrodensitometry.
    c. bioelectrical impedance.
    d. dual energy X-ray absorptiometry.

19. Overweight persons are faced with the possibility of these effects of obesity:
    1. earlier death due to a host of physical problems.
    2. precipitation of diabetes.
    3. lowered accident rate due to the protection from adipose tissue.
    4. increased risk of cardiovascular disease.
    5. greater economic success because they are left alone to concentrate on their jobs.

    a. 1, 3, 5
    b. 1, 2, 4
    c. 1, 2, 5
    d. 2, 3, 4
    e. 3, 4, 5

20. This risk factor is considered as strong as high blood cholesterol, hypertension and smoking for heart attack risk.

    a. overweight
    b. obesity
    c. central obesity
    d. cancer

ANSWERS

Summing Up—(1) kcalorie; (2) exchange; (3) heat; (4) body; (5) oxygen; (6) energy; (7) basal; (8) muscular; (9) lean; (10) hormonal; (11) 3,500; (12) two; (13) vitamins; (14) weight; (15) 20%; (16) excessive; (17) social; (18) underweight; (19) hereditary; (20) satiety; (21) inactivity.

Chapter study questions from the text—(1) Overfatness and underweight. (2) Hunger is the physiological need for food; appetite is the psychological desire for food. Satiation is the feeling of satisfaction and fullness that occurs during a meal and halts eating; satiety is a feeling of fullness after a meal. Hunger and appetite are likely to increase food intake. Satiation determines how much food is consumed during a meal. Satiety inhibits eating until the next meal; it determines how much time passes between meals. A high level of satiety is likely to decrease food intake between meals. (3) Basal metabolism: energy needed to maintain life when a body is at complete rest. Physical activity: voluntary movement of the skeletal muscles and support systems. Thermic effect of food: energy required to process food. Add the basal metabolism (1.0 kcal/kg for men; 0.9 kcal/kg for women X 24 hours/day) to the voluntary muscular activity expenditure (50-100% of the BMR depending on level of activity). (4) Weight depends on frame size which is difficult to measure; body composition is more important than weight but is difficult to measure. Location of body fat is more important than quantity of body fat, and assessment techniques include: fatfold measures, hydrodensitometry, and bioelectrical impedance analysis. (5) There is no standard weight against which weights can be compared. (6) Central obesity is excess fat on the abdomen and around the trunk of the body. It presents a greater risk to health than fat elsewhere on the body. (7) Increased risk of heart attacks, strokes, diabetes, high blood cholesterol, hypertension, surgery complications, gynecological problems, toxemia of pregnancy, certain types of cancer, arthritis, abdominal hernias, respiratory problems, gout, and accidents.

Short Answers—

1. basal metabolism; physical activities

Problem Solving—

1. 175 lb divided by 2.2 lb/kg = 79 kg
   79 kg x 1 kcal/kg/h = 79 kcal/hr
   79 kcal/h x 24 h/day = 1896 kcal/day

Crossword Puzzle—

Across: HYDRODENSITOMETRY; FATFOLDMEASURE; INDIRECT; BODYCOMPOSITION; BASALMETABOLISM

Down: LEANBODYMASS; THERMOGENESIS; HYPOTHALAMUS; DIRECT; COMI (COMPOSITION intersecting); BMI

Sample Test Questions—

1. d (p. 252)
2. a (p. 253)
3. c (p. 256)
4. c (p. 256)
5. c (p. 259)
6. b (p. 252)
7. b (p. 256)
8. b (p. 257)
9. d (p. 256)
10. d (p. 257)
11. a (p. 257)
12. d (p. 252)
13. c (p. 262)
14. b (p. 256)
15. a (p. 266)
16. a (p. 262)
17. a (p. 266)
18. b (p. 266)
19. b (p. 267)
20. c (p. 265)

CHAPTER 9
WEIGHT MANAGEMENT: OVERWEIGHT AND UNDERWEIGHT

CHAPTER OUTLINE

I. Overweight
    A. Fat Cell Development

    B. Fat Cell Metabolism

    C. Set-Point Theory

II. Causes of Obesity
    A. Genetics

    B. Environment

III. Problems of Obesity
    A. Health Risks

    B. Perceptions and Prejudices

    C.   Dangerous Interventions

IV.   Aggressive Treatments of Obesity
    A.   Drugs

    B.   Surgery

V.   Weight-Loss Strategies
    A.   Eating Plans

    B.   Physical Activity

    C.   Behavior and Attitude

    D.   Weight Maintenance

    E.   Prevention

    F.   Public Health Programs

VI. Underweight
    A. Problems of Underweight

_____
_____
_____

    B. Weight-Gain Strategies

_____
_____
_____

Highlight: Eating Disorders

_____
_____
_____
_____

SUMMING UP

Excess body fat accumulates when people consistently take in more food (1)_____ than they (2)_____. Obesity likely has many (3)_____. Probable causes of obesity include (4)_____ _____ development, genetics, fat cell (5)_____, (6)_____-_____ theory, overeating, and (7)_____. Since causes of obesity vary widely, treatment must be (8)_____ and multifaceted. Ineffective treatments include the use of (9)_____, amphetamines, and hormones. (10)_____ is useful only as a last resort.

The most important component of successful treatment is the adoption of a balanced and nourishing low-kcalorie diet; fat loss is greatly enhanced by regular (11)_____; and (12)_____ modification helps these adaptive strategies to be maintained. Major criteria for success are a permanent change in (13)_____ habits and (14)_____ of the goal weight over the long term.

Diet and behavior modification can also help the underweight person to gain weight. But the special case of anorexia (15)_____ requires skilled professional attention. In all weight-control problems—obesity, underweight, and anorexia—real success is achieved only when new, adaptive eating and coping (16)_____ have permanently replaced the old ones.

Chapter Glossary

bariatrics: field of medicine specializing in the treatment of obesity.

behavior modification: the changing of behavior by the manipulation of *antecedents* (cues, or environmental factors that trigger behavior), the *behavior* itself, and *consequences* (the penalties or rewards attached to behavior).

benzocaine: anesthetizes the tongue, reducing taste sensations.

brown adipose tissue: masses of specialized fat cells packed with pigmented mitochondria that produce heat instead of ATP.

cellulite: supposedly a lumpy form of fat; actually a fraud. The lumpy appearance in fatty areas of the body is caused by strands of connective tissue that attach the skin to underlying muscles. These points of attachment may pull tight where the fat is thick, making lumps appear between them. The fat itself is not different from fat anywhere else in the body. So, if the fat in these areas is lost, the lumpy appearance disappears.

clinically severe obesity: a BMI of 40 or greater or 100 pounds or more overweight for an average adult.

diuretic: a drug that promotes water excretion; popularly, a "water pill."

epidemic: the appearance of a disease or condition that attacks many people at the same time in the same region.

fad diets: popular eating plans that promise quick weight loss; most fad diets severely limit certain foods or overemphasize others.

gastric partitioning: a surgical procedure used to treat clinically severe obesity. The operation limits food intake by reducing the size of the stomach and delays gastric emptying by restricting the outlets.

gene pool: all the genetic information of a population at a given time.

ghrelin: a protein produced by the stomach cells that enhances appetite and decreases energy expenditure.

hyperplastic obesity: obesity due to an increase in the *number* of fat cells.

hypertrophic obesity: obesity due to an increase in the *size* of fat cells.

leptin: a protein produced by fat cells under direction of the *ob* gene that decreases appetite and increases energy expenditure.

lipotoxicity: adverse effects of fat in nonadipose tissues.

orlistat: a drug used in the treatment of obesity that inhibits the absorption of fat in the GI tract, thus limiting kcaloric intake: marketed under the trade name *Xenical*.

phenylpropanolamine: an ingredient commonly used to suppress appetite; reported side effects including dry mouth, rapid pulse, nervousness, sleeplessness, hypertension, irregular heartbeats, kidney failure, seizures, and strokes.

sibutramine: a drug used in the treatment of obesity that slows the reabsorption of serotonin in the brain, thus suppressing appetite and creating a feeling of fullness; marketed under the trade name *Meridia*.

serotonin: a neurotransmitter important in sleep regulation, appetite control and sensory perception.

set point: the point at which controls are set (for example, on a thermostat). The set point theory proposes that the body tends to maintain a certain weight by means of its own internal controls.

successful weight-loss maintenance: achieving a weight loss of at least 10 percent of initial body weight and maintaining the loss for at least one year.

toxic food environment: referring to easy access to and overabundance of high-fat, high-kcalorie foods in our society. It does not refer to the contamination of the food supply with poisonous toxins or infectious microbes.

weight cycling: repeated cycles of weight loss and gain, popularly called the *ratchet effect* or *yo-yo effect* of dieting.

ASSIGNMENTS

Answer these chapter study questions from the text:

1. Describe how body fat develops and suggest some reasons why it is difficult for an obese person to maintain weight loss.

2. What factors contribute to obesity?

3. List several aggressive ways to treat obesity and explain why such methods are not recommended for every overweight person.

4. Discuss reasonable dietary strategies suitable for achieving and maintaining a healthy body weight.

5. What are the benefits of increased physical activity in a weight-loss program?

6. Describe the behavioral strategies recommended for changing an individual's dietary habits. What role does personal attitude play?

7. Describe strategies for successful weight gain.

Complete this crossword puzzle by Mary A. Wyandt, Ph.D., CHES.

Across:	Down:
2. obesity in which the BMI is of 40 or greater or 100 pounds or more overweight for an average adult	1. a protein produced by the stomach cells that enhances appetite and decreases energy expenditure
4. all the genetic information of a population at a given time	3. adverse effects of fat in non-adipose tissues
7. masses of specialized fat cells packed with pigmented mitochondria that produce heat instead of ATP	5. an environment; referring to easy access to and overabundance of high-fat, high-kcalorie foods in our society
9. the point at which controls are set	6. supposedly a lumpy form of fat; actually a fraud
10. a protein produced by fat cells under direction of the ob gene that decreases appetite and increases energy expenditure	8. a neurotransmitter important in sleep regulation, appetite control and sensory perception

Sample Test Questions

1. Obesity due to an increase in the number of fat cells is called:

    a. hypertrophic obesity.
    b. hyperplastic obesity.
    c. epidemic obesity.
    d. none of the above.

2. This enzyme promotes efficient fat storage in both fat and muscle cells:

    a. protease.
    b. brown fat lipase.
    c. leptin protease.
    d. lipoprotein lipase.

3. The obesity gene is called:

    a. *ob* gene.
    b. set point.
    c. *leptin* gene.
    d. none of the above.

4. When leptin levels are high:

    a. the hypothalamus produces melanocortins.
    b. appetite is reduced.
    c. energy expenditure is slowed.
    d. a and b.
    e. a, b, and c.

5. Adverse reactions to fad diets can include:

    a. headaches and dizziness.
    b. nausea.
    c. death.
    d. a and b.
    e. a, b, and c.

6. A drug that slows the reabsorption of serotonin in the brain is called:

    a. orlistat.
    b. benzocaine.
    c. phenylpropanolamine.
    d. sibutramine.

7. Which of the following is accurate regarding chromium?

    a. It affects body composition and metabolism.
    b. It permits the achievement of higher levels of energy expenditure and thus weight loss.
    c. It is ineffective for eliminating body fat.
    d. It is a safe method of weight loss.

8. Methods to induce weight loss by interfering with the amount of food consumed include all of the following *except*

    a. liposuction.
    b. gastric partitioning.
    c. reducing the size of the stomach.
    d. restricting the outlet of the stomach.

9. Which of the following does *not* describe the controversies in obesity treatment?

    a. treating obesity is a simple task
    b. everyone cannot achieve thinness
    c. overweight people face discrimination
    d. self-esteem is harmed from repeated weight loss and gain

10. Only _____ percent of all people who lose weight maintain their losses for a year.

    a. 1
    b. 5
    c. 10
    d. 15

11. What is the best approach to weight loss?

    a. avoid foods containing carbohydrates
    b. reduce daily energy intake and increase energy expenditure
    c. eliminate all fats from the diet and decrease water intake
    d. greatly increase protein intake to prevent body protein loss

12. To lose fat efficiently while retaining lean tissue, a person needs a _____ kcalories deficit per day.

    a. 500
    b. 100
    c. 150
    d. 200

13. Fat intake is not important in a weight control program.

    a. true
    b. false

14. A tip for weight loss is:

    a. cut out all fats in the diet.
    b. increase your daily energy intake.
    c. avoid all foods containing carbohydrates.
    d. increase your water intake.

15. Regular physical activity often results in all of the following *except*:

    a. lower energy expenditure.
    b. raised BMR.
    c. improved appetite control.
    d. stress reduction.

16. Specific exercises can remove fat from certain targeted body parts.

    a. true
    b. false

17. To help maximize the long-term success of a person's weight-loss program, which of the following personal attitudes should be encouraged in the individual?

    a. Openness to examining emotional health status and whether stress triggers overeating.
    b. Viewing the body realistically as being very fat rather than thin.
    c. Refraining from expressing overconfidence in ability to lose weight.
    d. Accepting that lack of exercising is a part of the lifestyle of most overweight people.

18. All of the following are behavior modifications for losing weight *except*

    a. shopping only when not hungry.
    b. eating only in one place and in one room.
    c. watching television viewing only when not eating.
    d. taking smaller portions of food but eating everything quickly.

19. Which of the following would *not* be part of a successful program of weight gain in an underweight individual?

    a. engage in physical exercise to build muscle tissue
    b. consume energy-dense foods
    c. consume energy-dense beverages
    d. consume a large number of small meals

20. It takes an excess of approximately _____ kcalories per day above normal energy needs to support the gain of pure lean tissue.

    a. 300
    b. 500
    c. 550-600
    d. 750-800

ANSWERS

Summing Up—(1) energy; (2) spend; (3) interrelated; (4) fat cell; (5) metabolism; (6) set-point; (7) inactivity; (8) individualized; (9) diuretics; (10) Surgery; (11) exercise; (12) behavior; (13) eating; (14) maintenance; (15) nervosa; (16) behaviors.

Chapter study questions from the text—(1) Body fat develops when fat cells increase in number and size. Prevention of excess weight gain depends on maintaining a reasonable number of fat cells; when an obese person loses weight, the body attempts to return to the original weight, or its set point. (2) Genetics (leptin, ghrelin, uncoupling problems, fat cell metabolism, set point), overeating, inactivity. (3) Fad diets, diuretics, amphetamines, other prescription drugs, over-the-counter drugs, hot baths, machines that jiggle, brushes, sponges, massages, surgery, gastric partitioning, gastric balloons, jaw wiring, very-low-kcalorie diets. Reasons these are not recommended for everyone: weight cycling, psychology of weight cycling, some methods do not work at all, others only provide temporary weight loss (not fat loss). (4) Eating plans based on realistic energy intake, nutritional adequacy, physical activity, making small behavior modification changes, support groups. (5) Physical activity increases BMR, helps control appetite, provides psychological benefits. (6) Record food intake using a food record to help become aware of behaviors, focus on learning desired eating and exercise behaviors and eliminating unwanted behaviors, do not grocery shop when hungry, eat slowly, exercise when watching television. A personal attitude of sound emotional health supports positive behavior change. (7) Eat energy-dense foods to provide excess of 700-1000 kcals per day, eat regular meals daily, eat large portions, eat extra snacks between meals, drink plenty of juice and milk, exercise to build muscle.

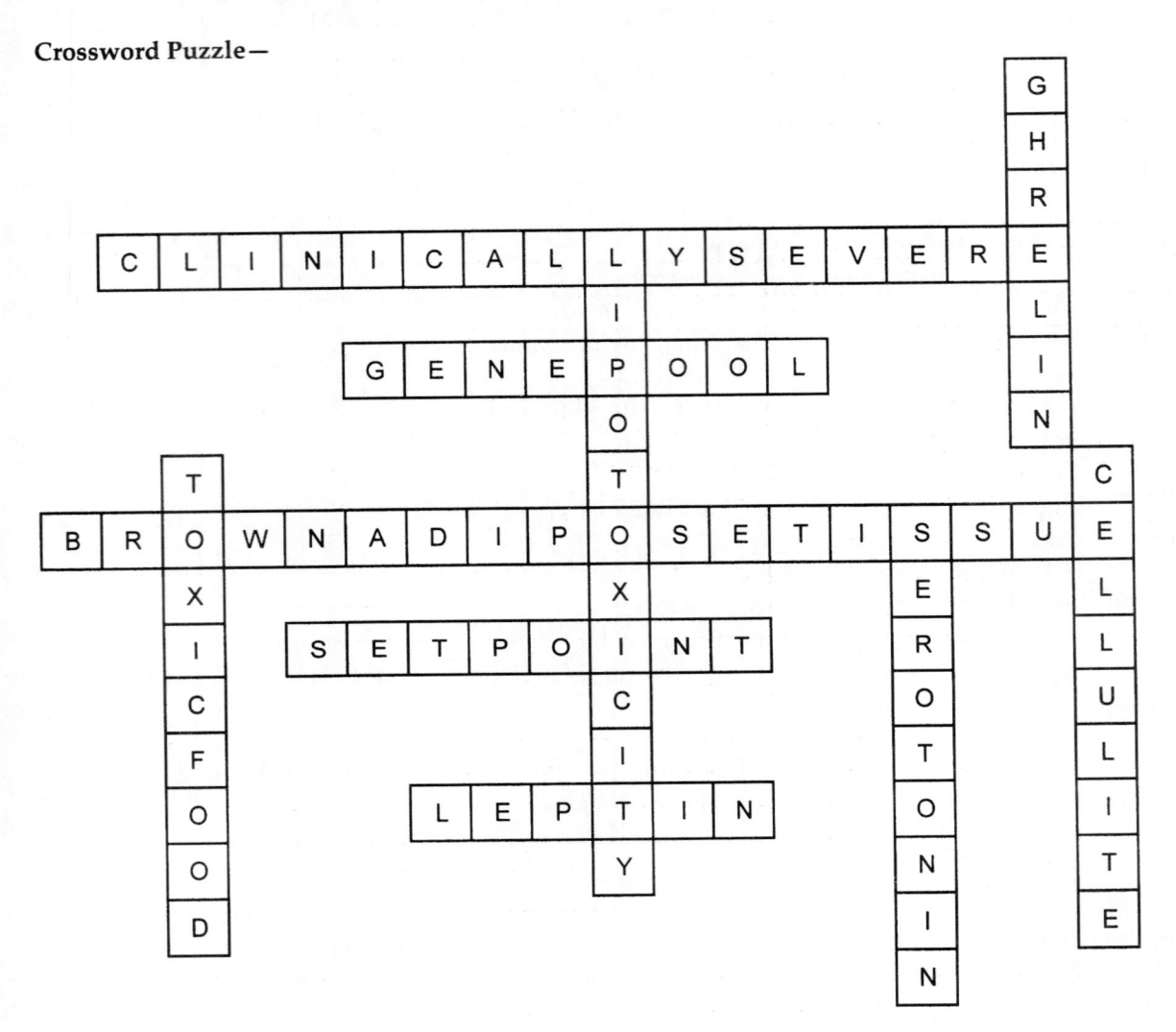

Crossword Puzzle—

Across/Down solution:
- CLINICALLYSEVERE
- GHRELIN
- GENEPOOL
- BROWNADIPOSETISSUE
- TOXICFOOD
- SETPOINT
- LEPTIN
- SEROTONIN
- CELLULITE
- LIPOPROTEINLIPASE
- OXYCIN

Sample Test Questions—

1.	b (p. 281)	8.	a (p. 290)	15.	a (p. 296)
2.	d (p. 280)	9.	a (p. 286, 287)	16.	b (p. 298)
3.	a (p. 282)	10.	b (p. 300)	17.	a (p. 299)
4.	d (p. 282)	11.	b (p. 299, 300)	18.	d (p. 298)
5.	e (p. 287)	12.	a (p. 293)	19.	d (p. 302, 303)
6.	d (p. 290)	13.	b (p. 295)	20.	d (p. 302)
7.	c (p. 289)	14.	d (p. 294)		

Chapter 10
The Water-Soluble Vitamins: B Vitamins and Vitamin C

Chapter Outline

I. The Vitamins—An Overview

II. The B Vitamins—As Individuals
   A. Thiamin

   B. Riboflavin

   C. Niacin

   D. Biotin

   E. Pantothenic Acid

   F. Vitamin B$_6$

   G. Folate

H.   Vitamin B$_{12}$

   I.   Non-B Vitamins

III.   The B Vitamins—In Concert
   A.   B Vitamin Roles

   B.   B Vitamin Deficiencies

   C.   B Vitamin Toxicities

   D.   B Vitamin Food Sources

IV.   Vitamin C
   A.   Vitamin C Roles

   B.   Vitamin C Recommendations

   C.   Vitamin C Deficiency

D.   Vitamin C Toxicity

E.   Vitamin C Food Sources

Highlight: Vitamin and Mineral Supplements

SUMMING UP

The B vitamins serve as (1)_____ assisting many enzymes in the body. Thiamin, (2)_____, niacin, and pantothenic acid are especially important in the glucose-to-energy pathway; they are active in the coenzymes (3)_____, FAD, NAD+, and (4)_____, respectively.

Vitamin B₆ facilitates (5)_____ acid transformations and thus protein metabolism; (6)_____ is involved in pathways leading to the synthesis of new cells, and (7)_____ in the release of folate in its active form; and (8)_____ is involved in lipid synthesis.

A lack of thiamin causes (9)_____; a lack of niacin (unless compensated for by its amino acid precursor tryptophan) causes (10)_____; a lack of the intrinsic factor for vitamin B₁₂ causes (11)_____ anemia. Human deficiencies have also been observed for riboflavin, vitamin B₆, and folate.

(12)_____ is widely distributed in foods, but no food contributes a great amount of it. (13)_____ is concentrated in milk and meats. (14)_____ is found wherever protein is found and can also be made from the amino acid tryptophan. Vitamin B₆ is most abundant in (15)_____, vitamin B₁₂ is found only in (16)_____ products, and (17)_____ is supplied best by green, leafy vegetables. Any diet plan that includes moderate amounts of all these foods ensures probable (18)_____ for these nutrients.

Vitamin C acts as an (19)_____, helping to maintain iron in its reduced (20)_____ form and thus cooperating with enzymes that require this form of iron as a (21)_____.

Vitamin C also helps regulate the overall oxidation-reduction equilibrium of the body (22)_____ and fluids. Vitamin C promotes the formation of the protein (23)_____. It is involved in the metabolism of several (24)_____ _____. Deficiency of vitamin C causes (25)_____. Scurvy is prevented by the daily intake of only (26)_____ milligrams of vitamin C. Recommended daily intakes of vitamin C range from (27)_____ to _____ milligrams. Toxic effects of megadoses (3 to 10 grams) have been reported. The best food sources of vitamin C are the (28)_____ fruits, strawberries and cantaloupe, broccoli and other members of the cabbage family, and greens.

CHAPTER GLOSSARY

anemia: literally, "too little blood"; any condition in which too few red blood cells are present, or the red blood cells are immature (and therefore large) or too small or contain too little hemoglobin to carry the normal amount of oxygen to the tissues. It is not a disease itself, but can be a symptom of many different disease conditions.
antagonist: a competing factor that counteracts the action of another factor; when a drug displaces a vitamin from its site of action, the drug renders the vitamin ineffective and thus acts as a vitamin antagonist.
antioxidant: a substance in foods that significantly decreases the adverse effects of free radicals on normal physiological functions in the human body.
antiscorbutic factor: the original name for vitamin C.
ariboflavinosis: riboflavin deficiency causing inflammation of the membranes of the mouth, skin, eyes, and GI tract.
ascorbic acid: one of the two active forms of vitamin C.
atrophic gastritis: chronic inflammation of the stomach accompanied by a diminished size and functioning of the mucous membranes and glands.
avidin: the protein in egg whites that binds biotin.
beriberi: the thiamin-deficiency disease.
bioavailability: the rate at and the extent to which a nutrient is absorbed and used.
biotin: a B vitamin that functions as a coenzyme in the metabolism of carbohydrates and fats.
carnitine: a nonessential nutrient made in the body from the amino acid lysine.
carpal tunnel syndrome: a pinched nerve at the wrist, causing pain or numbness in the hand; it is often caused by repetitive motions of the wrist.
cheilosis: a symptom of vitamin B deficiencies; a condition of reddened lips with cracks at the corners.
choline: a nitrogen-containing compound found in foods and made in the body from an amino acid. Choline is used in the body to make the phospholipid lecithin and the neurotransmitter acetylcholine.
cofactor: a small inorganic or organic substance that works with an enzyme to facilitate a chemical reaction.

dietary folate equivalents (DFE): the amount of folate available to the body from naturally occurring sources, fortified foods, and supplements, accounting for differences in the bioavailability from each source.

false negative: a test result indicating that a condition is *not* present (negative) when in fact it is present (therefore false).

false positive: a test result indicating that a condition is present (positive) when in fact it is not (therefore false).

folate: a B vitamin; also known as folic acid, folacin, or pteroylglutamic acid (PGA); the coenzyme forms are DHF (dihydrofolate) and THF (tetrahydrofolate).

glossitis: a symptom of vitamin B deficiencies, an inflammation of the tongue.

gluconeogenesis: the synthesis of glucose from noncarbohydrate sources such as amino acids or glycerol.

histamine: a substance produced by cells of the immune system as part of a local immune reaction to an antigen; participates in causing inflammation.

inositol: a nonessential nutrient that can be made in the body from glucose. Inositol is used in cell membranes.

intrinsic factors: a glycoprotein (a protein with short polysaccharide chains attached) made in the stomach that aids in the absorption of vitamin B_{12}.

macrocytic or megaloblastic anemia: the large-cell anemia.

neural tube defects: malformations of the brain, spinal cord, or both during embryonic development.

niacin: a B vitamin. Niacin can be eaten preformed or can be made in the body from its precursor, tryptophan, one of the amino acids. The coenzyme forms are NAD (nicotinamide adenine dinucleotide) and NADP (the phosphate form of NAD).

niacin equivalents (NE): the amount of niacin present in food, including the niacin that can theoretically be made from its precursor, tryptophan, present in the food.

niacin flush: a temporary burning, tingling, and itching sensation that occurs when a person takes a large dose of nicotinic acid; often accompanies by a headache and reddened face, arms and chest.

nutritional yeast: a preparation of yeast cells grown especially as a nutrient supplement, particularly for vegetarian diets.

oxidative stress: an imbalance between the production of free radicals and the body's ability to handle them and prevent damage.

pantothenic acid: a B vitamin; the principal active form is part of coenzyme A, called "CoA."

pellagra: the niacin-deficiency disease causing dermatitis.

pernicious anemia: a blood disorder that reflects a vitamin B_{12} deficiency caused by lack of intrinsic factor and characterized by abnormally large and immature red blood cells. Other symptoms include muscle weakness, and neurological disturbances.

pharmacological effect: when a large dose (two to ten times greater than the RDA) overwhelms some body system and acts like a drug.

physiological effect: when a normal dose of a nutrient (levels commonly found in foods and not exceeding 150% of the RDA) provides a normal blood concentration.

precursors: substances that precede others; with regard to vitamins, compounds that can be converted into active vitamins; also known as *provitamins*.

riboflavin: a B vitamin; the coenzyme forms are FMN (flavin mononucleotide) and FAD (flavin adenine dinucleotide).

scurvy: the vitamin C-deficiency disease.

serotonin: a neurotransmitter important in appetite control, sleep regulation, and sensory perception; it is synthesized from the amino acid tryptophan with the help of vitamin B_6.

thiamin: a B vitamin; the coenzyme form is TPP (thiamin pyrophosphate).

vitamin B₆: a family of compounds—pyridoxal, pyridoxine, and pyridoxamine; the primary active coenzyme form is PLP (pyridoxal phosphate).

vitamin B₁₂: a B vitamin characterized by the presence of cobalt; the active forms of coenzyme B₁₂ are methylcobalamin and deoxyadenosylcobalamin.

Wernicke-Korsakoff Syndrome: severe thiamin deficiency in alcohol abusers; symptoms include disorientation, loss of short-term memory, jerky eye movements, and staggering gait.

Assignments

Answer these chapter study questions from the text:

1. How do the vitamins differ from the energy nutrients?

2. Describe some general differences between fat-soluble and water-soluble vitamins.

3. Which B vitamins are involved in energy metabolism? Protein metabolism? Cell division?

4. For thiamin, riboflavin, niacin, biotin, pantothenic acid, vitamin B₆, folate, vitamin B₁₂, and vitamin C, state:

   - Its chief function in the body.
   - Its characteristic deficiency symptoms.
   - Its significant food sources.

5. What is the relationship of tryptophan to niacin?

6. Describe the relationship between folate and vitamin B$_{12}$.

7. What risks are associated with high doses of niacin? Vitamin B$_6$? Vitamin C?

Complete these short answer questions:

1. Characteristics of water-soluble vitamins are:

    a.   c.

    b.   d.

2. The names of the water-soluble vitamins are:

    a.   f.

    b.   g.

    c.   h.

    d.   i.

    e.

3. Distinguish between these types of deficiencies:

    a.   primary:

    b.   secondary:

Solve these problems:

1. A person whose RDA for protein is 45 grams consumes 75 grams of protein in a day. How many niacin equivalents does this represent?

2. What percentage of the adult RDA for vitamin C does a person receive from 1/2 cup of broccoli?

Complete this crossword puzzle by Mary A. Wyandt, Ph.D., CHES.

	Across:	Down:
2.	a substance in foods that significantly decreases the adverse effects of free radicals on normal physiological functions in the human body	1. a B vitamin; the coenzyme forms are FMN and FAD
6.	a nitrogen-containing compound found in foods and made in the body from an amino acid	3. a B vitamin; the coenzyme form is TPP
		4. a nonessential nutrient that can be made in the body from glucose; it is used in cell membranes
8.	a B vitamin that functions as a coenzyme in the metabolism of carbohydrates and fats	5. a nonessential nutrient made in the body from the amino acid lysine
9.	a B vitamin, which can be eaten preformed or can be made in the body from its precursor, tryptophan	7. the protein in egg whites that binds biotin
10.	a B vitamin; also known as folic acid	

SAMPLE TEST QUESTIONS

1. Vitamins are:

    a. organic.
    b. inorganic.
    c. essential nutrients required in small amounts.
    d. a and c
    e. b and c

124

2. Vitamins differ from carbohydrates, protein, and fat in the following way(s):

   a. function.
   b. structure.
   c. amounts required.
   d. a and b only
   e. a, b, and c

3. Water-soluble vitamins consumed in excess of need may be:

   a. malabsorbed.
   b. beneficial.
   c. harmful.
   d. a and b
   e. a and c

4. B vitamins:

   a. are water soluble.
   b. function as coenzymes.
   c. are stored in the adipose tissue.
   d. are absorbed into the lymph.
   e. a and b

5. A dietary deficiency of B vitamins can cause:

   a. negative nitrogen balance.
   b. impairment of energy metabolism.
   c. impaired iron utilization.
   d. reduced thyroid production.

6. A deficiency of thiamin produces the disease:

   a. rickets.
   b. pellagra.
   c. beriberi.
   d. scurvy.

7. Which of the following vitamins is most readily destroyed by ultraviolet light?

   a. niacin
   b. thiamin
   c. riboflavin
   d. ascorbic acid

8. The dietary need for _____ is influenced by the presence of the amino acid tryptophan in the diet.

   a. thiamin
   b. pantothenic acid
   c. niacin
   d. biotin
   e. riboflavin

9. Two coenzyme forms of niacin are:

   a. FMN and FAD.
   b. NAD and FAD.
   c. NADP and TPP.
   d. FAD and TPP.
   e. NADP and NAD.

10. The disease called pellagra is related to a:

    a. dietary deficiency of riboflavin.
    b. low protein diet high in corn products.
    c. diet high in polished rice.
    d. a and b
    e. a and c

11. Which of the following is highest in riboflavin content per serving?

    a. butter
    b. milk
    c. bread
    d. apple
    e. egg

12. If your skin feels flushed after taking a vitamin pill you may have overdosed on:

    a. niacin.
    b. thiamin.
    c. riboflavin.
    d. folate.

13. Pantothenic acid is a part of the structure of:

    a. pyruvate.
    b. glucose.
    c. cobalamin.
    d. glycogen.
    e. coenzyme A.

14. Often a function of one vitamin depends on the presence of another. This interdependence is shown in these 2 vitamins.

    a. vitamin B₁₂ and folate
    b. vitamin A and vitamin C
    c. vitamin B₁₂ and niacin
    d. folate and vitamin C

15. Vitamin B₁₂ is different from other B vitamins because:

    a. it may be synthesized from a certain amino acid.
    b. it requires a carrier to be transported from the intestinal tract to the bloodstream.
    c. deficiencies of vitamin B₁₂ never occur in human beings.
    d. it is found only in plant foods.

16. _____ may be lacking in the diet of strict vegetarians.

    a. Vitamin B₁₂
    b. Thiamin
    c. Riboflavin
    d. Niacin
    e. Biotin

17. The best food source of folate, among the following, is:

    a. spinach.
    b. milk.
    c. coffee.
    d. ice cream.

18. The vitamin C deficiency disease is:

    a. ascorbic acidosis.
    b. scurvy.
    c. pellagra.
    d. beriberi.

19. Most of the symptoms of a vitamin C deficiency are caused by:

    a. anemia.
    b. failure to maintain integrity of blood vessels.
    c. inactivity of intestinal bacteria.
    d. decreased utilization of protein.

20. The amount of vitamin C needed to prevent scurvy is about:

    a. 10 milligrams a day.
    b. 35 milligrams a day.
    c. 60 milligrams a day.
    d. 100 milligrams a day.

21. Which of the following is lowest in vitamin C?

    a. strawberries
    b. potatoes
    c. milk
    d. cauliflower
    e. orange juice

ANSWERS

Summing Up—(1) coenzymes; (2) riboflavin; (3) TPP; (4) coA; (5) amino; (6) folate; (7) vitamin B$_{12}$; (8) biotin; (9) beriberi; (10) pellagra; (11) pernicious; (12) Thiamin; (13) Riboflavin; (14) Niacin; (15) meats; (16) animal; (17) folate; (18) adequacy; (19) antioxidant; (20) (ferrous iron); (21) cofactor; (22) cells; (23) collagen; (24) amino acids; (25) scurvy; (26) 10; (27) 30 to 75; (28) citrus.

Chapter study questions from the text—(1) They differ in structure, function, and food contents. (2) Water-soluble vitamins are: carried in the blood, excreted in the urine, needed in frequent, small doses, and unlikely to reach toxic levels in the body. Fat-soluble vitamins are: absorbed into the lymph and carried in the blood by protein carriers, stored in body fat, needed in periodic doses, and more likely to be toxic when consumed in excess of needs. (3) B vitamins involved in energy metabolism: thiamin, riboflavin, niacin, biotin, pantothenic acid. B vitamin involved in protein metabolism: vitamin B$_6$. B vitamins involved in cell division: folate and vitamin B$_{12}$. (4) See text for respective summary tables for each nutrient. (5) Tryptophan can be converted to niacin in the body: 60 mg tryptophan = 1 mg niacin. (6) Vitamin B$_{12}$ and folate's roles intertwine because each depends on the other for activation. Folate is part of a coenzyme that helps convert vitamin B$_{12}$ to one of its coenzyme forms and helps synthesize DNA; vitamin B$_{12}$ removes a methyl group to activate the folate coenzyme. (7) Niacin: diarrhea, heartburn, nausea, ulcer irritation, vomiting, fainting, dizziness, painful flush and rash, excessive sweating, liver damage, low blood pressure. Vitamin B$_6$: depression, fatigue, headaches, nerve damage, muscle weakness, numbness, bone pain. Vitamin C: nausea, abdominal cramps, diarrhea, false results in diabetes tests, interference in drug effectiveness, hemolytic anemia, stone formation, rebound scurvy.

Short Answers—
1. carried in bloodstream; excreted in urine; needed in frequent small doses; less likely to be toxic
2. thiamin, riboflavin, vitamin B6, niacin, folate, vitamin B12, biotin, pantothenic acid, vitamin C
3. primary = inadequate intake; secondary = impaired absorption or metabolism, or excessive excretion

Problem Solving—
1. 75 g - 45 g = 30 g protein
   30 g divided by 100 = .3 g tryptophan
   .3 g X 1000 = 300 mg tryptophan
   300 mg divided by 60 = 5 mg niacin equivalents
2. 98 mg/c X .5 c = 49 mg
   49 mg divided by 60 mg (RDA) X 100 = 82% (rounded)

Crossword Puzzle—

```
            R
A N T I O X I D A N T
            B
            O
        T   F       I           C
        C H O L I N E           A
    A   I   A       O           R
    V   A   V       S           N
    I   M   I       I           I
    D   I   N       T   B I O T I N
N I A C I N         O           N
    N               F O L A T E
```

Sample Test Questions—
1. d (p. 321)
2. e (p. 322)
3. c (p. 323)
4. e (p. 324, 325)
5. b (p. 325)
6. c (p. 326)
7. c (p. 328)
8. c (p. 330)
9. e (p. 330)
10. b (p. 330)
11. b (p. 329)
12. a (p. 330)
13. e (p. 333)
14. a (p. 335)
15. b (p. 340)
16. a (p. 342)
17. a (p. 340)
18. b (p. 351)
19. b (p. 351)
20. a (p. 350)
21. c (p. 352)

Chapter 11
The Fat-Soluble Vitamins: A, D, E, and K

Chapter Outline

I. Vitamin A and Beta-Carotene
    A. Roles in the Body

    B. Vitamin A Deficiency

    C. Vitamin A Toxicity

    D. Vitamin A Recommendations

    E. Vitamin A in Foods

II. Vitamin D
    A. Roles in the Body

    B. Vitamin D Deficiency

    C. Vitamin D Toxicity

    D.   Vitamin D Recommendations and Sources

III.   Vitamin E
    A.   Vitamin E as an Antioxidant

    B.   Vitamin E Deficiency

    C.   Vitamin E Toxicity

    D.   Vitamin E Recommendations

    E   Vitamin E in Foods

IV.   Vitamin K
    A.   Roles in the Body

    B.   Vitamin K Deficiency

    C.   Vitamin K Toxicity

D. Vitamin K Recommendations and Sources

V. The Fat-Soluble Vitamins—In Summary

Highlight: Antioxidant Nutrients in Disease Prevention

SUMMING UP

Vitamin A is a part of visual (1)_____ and is essential for vision. It is involved in maintaining mucous (2)_____, helps maintain the skin, and is essential for the remodeling of (3)_____ during growth or mending. It plays a part in cell membranes, in (4)_____ synthesis, and in reproduction. Deficiency of vitamin A causes (5)_____ blindness; disorders of the respiratory, urogenital, reproductive, and nervous systems; and abnormalities of bones and teeth. (6)_____ symptoms are caused by large excesses (ten times the recommended intake or more) taken over a prolonged period and result only from the (7)_____ vitamin (from (8)_____ or animal products such as liver)—not from the precursor (9)_____ and its relatives.

Vitamin D promotes intestinal (10)_____ of calcium, mobilization of calcium from (11)_____ stores, retention of calcium by the (12)_____, and is therefore essential for the calcification of bones and teeth. Given reasonable exposure to (13)_____, human beings can synthesize this vitamin in the (14)_____. Deficiency of vitamin D causes rickets in children and (15)_____ in adults; excesses cause abnormally high blood calcium levels and result in deposition of calcium crystals in soft tissues, such as the kidneys and major blood vessels.

The most substantiated role of vitamin E in human beings is as an (16)_____ that protects vitamin A and the polyunsaturated fatty acids (PUFA) from destruction by oxygen. Vitamin E also protects the lungs against oxidizing air (17)_____. Only one vitamin E deficiency has been confirmed in human beings: erythrocyte (18)_____.

Vitamin K, the coagulation vitamin, promotes normal blood (19)_____; deficiency causes hemorrhagic disease. The vitamin is synthesized by intestinal (20)_____ and is available from foods such as green vegetables and milk. Deficiency is normally seen only in (21)_____, whose intestinal flora have not become established, in people taking (22)_____ or sulfa drugs, or in people whose fat absorption is impaired.

CHAPTER GLOSSARY

acne: a chronic inflammation of the skin's follicles and oil-producing glands, which leads to an accumulation of oils inside the ducts that surround hairs; usually associated with the maturation of young adults.
alpha-tocopherol: the active vitamin E compound.
beta-carotene: one of the carotenoids; an orange pigment and vitamin A precursor found in plants.
carotene: a vitamin A precursor found in plants; an orange pigment.
carotenoids: pigments commonly found in plants and animals, some of which have vitamin A activity.
cell differentiation: the process by which immature cells develop specific functions different from those of the original that are characteristic of their mature cell type.
chlorophyll: the green pigment of plants, which absorbs photons and transfers their energy to other molecules, thereby initiating photosynthesis.
cornea: the transparent membrane covering the outside of the eye.
D,L: D stands for *dextro*, or "right-handed," and L, for *levo*, or "left-handed," referring to the shapes of the molecules, which are mirror images of each other.
differentiation: development of specific functions different from those of the original.
epithelial cells: cells on the surface of the skin and mucous membranes.
epithelial tissues: the layer of the body that serves as a selective barrier between the body's interior and the environment.
ergocalciferol: a plant version of vitamin D, (also called vitamin D₂).
erythrocyte: red blood cell.
erythrocyte hemolysis: the breaking open of red blood cells; a symptom of vitamin E-deficiency disease in human beings.
fibrocystic breast disease: a harmless condition in which the breasts develop lumps, sometimes associated with caffeine consumption.
hair follicle: a group of cells in the skin from which a hair grows.
hemolysis: bursting of red blood cells.
hemolytic anemia: the condition of having too few red blood cells as a result of erythrocyte hemolysis.
hemophilia: a hereditary disease that is caused by a genetic defect but has no relation to vitamin K; the blood is unable to clot because it lacks the ability to synthesize certain clotting factors.
hemorrhagic disease: a disease characterized by excessive bleeding.
hypercalcemia: high blood calcium that may develop from a variety of disorders, including vitamin D toxicity. It does *not* develop from high calcium intake.

intermittent claudication: severe calf pain caused by inadequate blood supply; it occurs when walking and subsides during rest.
international units (IU): a measure of vitamin activity, determined by such biological methods as feeding a compound to vitamin-deprived animals and measuring growth.
jaundice: yellowing of the skin, due to spillover of the bile pigments bilirubin from the liver into the general circulation; also known as *hyperbilirubinemia*.
keratin: a water-insoluble protein; the normal protein of hair and nails.
keratinization: accumulation of keratin in a tissue; a sign of vitamin A deficiency.
keratomalacia: softening of the cornea that leads to irreversible blindness seen in severe vitamin A deficiency.
lysosomes: sacs of degradative enzymes.
mendione: a synthetic form of vitamin K.
mineralization: the process in which calcium, phosphorus, and other minerals crystallize on the collagen matrix of a growing bone, hardening the bone.
mucous membranes: the membranes, composed of mucus-secreting cells, that line the surfaces of body tissues.
mucus (adjective mucous): a class of substances secreted by the goblet cells of mucous membranes; a glycoprotein
muscular dystrophy: a hereditary disease in which the muscles gradually weaken; its most debilitating effects arise in the lungs.
night blindness: slow recovery of vision after flashes of bright light at night or an inability to see in dim light; an early symptom of vitamin A deficiency.
opsin: the protein portion of the visual pigment molecule.
osteoblasts: cells that build bone.
osteoclasts: the cells that destroy bone during growth.
osteomalacia: a bone disease characterized by softening of the bones; symptoms include bending of the spine and bowing of the legs.
phylloquinone: the natural form of vitamin K.
pigment: a molecule capable of absorbing certain wavelengths of light so that it reflects only those that we perceive as a certain color.
preformed vitamin A: dietary vitamin A in its active form.
RAE (retinol activity equivalent): a measure of vitamin A activity; the amount of retinol that the body will derive from a food containing preformed retinol or its precursor beta-carotene.
remodeling: the dismantling and reformation of a structure, in this case, bone.
retina: the layer of light-sensitive nerve cells lining the back of the inside of the eye; consists of rods and cones.
retinal: the aldehyde form of vitamin A, active in the eye.
retinoic acid: the acid form of vitamin A.
retinoids: chemically related compounds with biological activity similar to that of retinol.
retinol: the alcohol form of vitamin A.
retinol-binding protein (RBP): the specific protein responsible for transporting retinol.
rhodopsin: a light-senstive pigment of the retina. It contains the retinal form of vitamin A and the protein opsin.
rickets: the vitamin D-deficiency disease in children characterized by inadequate mineralization of bone.
sterile: free of microorganisms, such as bacteria.
teratogenic: causing abnormal fetal development and birth defects.
tocopherol: a general term for several chemically related compounds, most of which have vitamin E activity.

tocopherol equivalents (TE): the units in which vitamin E activity is measured. One TE equals the amount of vitamin E activity in 1 milligram of D-alpha-tocopherol.
tocotrienol: less active forms of vitamin E.
urethra: the tube through which urine from the bladder passes out of the body.
vitamin A: all naturally occurring compounds with the biological activity of retinal, the alcohol form of vitamin A.
vitamin A activity: a term referring to both the active forms of vitamin A and the precursor forms in foods without distinguishing between them.
vitamin D-refractory rickets: a rare type of rickets, *not* caused by vitamin D deficiency.
xanthophylls: pigments found in plants; responsible for the color changes seen in autumn leaves.
xerophthalmia: progressive blindness caused by vitamin A deficiency.
xerosis: abnormal drying of the skin and mucous membranes; a sign of vitamin A deficiency.

ASSIGNMENTS

Answer these chapter study questions from the text:

1. List the fat-soluble vitamins. What characteristics do they have in common? How do they differ from the water-soluble vitamins?

2. Summarize the roles of vitamin A and the symptoms of its deficiency.

3. What is meant by vitamin precursors? Name the precursors of vitamin A, and tell in what classes of foods they are located. Give examples of foods with high vitamin A activity.

4. How is vitamin D unique among the vitamins? What is its chief function? What are the richest sources of this vitamin?

5. Describe vitamin E's role as an antioxidant. What are the chief symptoms of vitamin E deficiency?

6. What is vitamin K's primary role in the body? What conditions may lead to vitamin K deficiency?

Complete these short answer questions:

1. Characteristics of fat-soluble vitamins are:

    a.   c.

    b.   d.

2. The names of the fat-soluble vitamins are:

    a.   c.

    b.   d.

Solve these problems:

1. If a carrot has 7,930 IU of vitamin A, approximately how many RE does it provide? If a piece of liver has 45,390 IU of vitamin A, approximately how many RE does it provide?

2. What percentage of the U.S. RDA for vitamin A does a person receive from 1 oz of liver?

Complete this crossword puzzle by Mary A. Wyandt, Ph.D., CHES.

Across:	Down:
3. a water-insoluble protein; the normal protein of hair and nails	1. a synthetic form of vitamin K
6. a vitamin A precursor found in plants; an orange pigment	2. pigments found in plants; responsible for the color changes seen in autumn leaves
8. the green pigment of plants, which absorbs photons and transfers their energy to other molecules	4. a plant version of vitamin D (a.k.a. vitamin D_2)
9. the protein portion of the visual pigment molecule	5. the alcohol form of vitamin A
10. less active forms of vitamin E	7. red blood cell

SAMPLE TEST QUESTIONS

1. A characteristic of the fat-soluble vitamins is that excesses are stored in the liver and fatty tissues.

    a. True
    b. False

2. Retinal is the _____ form of vitamin A.

    a. alcohol
    b. aldehyde
    c. acid
    d. provitamin A carotenoid

3. Night vision is maintained by:

    a. vitamin A, which combines with opsin in the dark to regenerate rhodopsin.
    b. vitamin D, which combines with opsin in the dark to regenerate rhodopsin.
    c. vitamin A, which combines with rhodopsin in the dark to regenerate retinene and opsin.
    d. vitamin E, which protects the polyunsaturated fatty acids in the membranes of the rod cells.

4. Which of the following surfaces is (are) not lined with epithelial cells?

    a. bladder and urethra
    b. mouth, stomach, and intestines
    c. eyelids
    d. lungs
    e. bones

5. This vitamin is needed for bones to grow in length.

    a. vitamin A
    b. vitamin B$_1$
    c. vitamin C
    d. vitamin D

6. An early sign of vitamin A deficiency is:

    a. rickets.
    b. hemolytic anemia.
    c. night blindness.
    d. total blindness.

7. Which of the following is a symptom of a vitamin A deficiency?

    a. anemia
    b. fissuring at the corners of the mouth
    c. keratinization of the epithelial cells
    d. erythematous areas closely resembling sunburn appearing on the skin

8. Vitamin A toxicity:

    a. can pose a teratogenic risk.
    b. is impossible.
    c. is helpful for clearing up acne.
    d. none of the above.

9. Among fruits and vegetables, the best sources of vitamin A are:

   a. green or yellow, such as lettuce and corn.
   b. dark green or deep orange, such as broccoli and sweet potatoes.
   c. green, such as lettuce, peas and snap beans.
   d. brightly colored, such as tomatoes and lemons.

10. Ultraviolet rays from the sun allow this vitamin to be synthesized.

    a. vitamin A
    b. vitamin B$_{12}$
    c. vitamin C
    d. vitamin D

11. Which of the following statements is true about vitamin D?

    a. It must be provided in the diet.
    b. It is an inorganic compound.
    c. It can be made in the body.
    d. It is not needed in adulthood.

12. The most important physiological function of vitamin D is:

    a. synthesis of red blood cells.
    b. promotion of calcium and phosphorus utilization.
    c. increased resistance to disease.
    d. prevention of night blindness.

13. The vitamin D-deficiency disease of children is:

    a. xerophthalmia.
    b. night blindness.
    c. follicular hyperkeratosis.
    d. rickets.

14. The major role of vitamin E in the body is to:

    a. aid in normal blood clotting.
    b. act as an antioxidant.
    c. aid in formation of normal epithelial tissue.
    d. aid in protein metabolism.

15. The one symptom of vitamin E deficiency that has been demonstrated in human beings is:

    a. weak bones.
    b. muscle paralysis.
    c. reproductive failure.
    d. breakage of red blood cell membranes.

16. Vitamin E supplements may be beneficial to:

    a. people with intermittent claudication.
    b. reverse damage caused by atherosclerosis.
    c. women with "hot flashes."
    d bladder cancer patients.

17. Vitamin E supplementation can cure hereditary muscular dystrophy in humans.

    a. True   b. False

18. Vitamin E need varies with a person's intake of:

    a. polyunsaturated fatty acids.   c. cholesterol.
    b. saturated fatty acids.   d. other fat soluble vitamins.

19. Vitamin K is necessary for:

    a. normal vision.   c. normal muscle growth.
    b. normal blood clotting.   d. prevention of night blindness.

20. Some of our vitamin K requirement is met by:

    a. synthesis of the vitamin by intestinal bacteria.
    b. synthesis of the vitamin from sunlight.
    c. synthesis of the vitamin from carotene.
    d. fortification of milk.

Answers

Summing Up—(1) pigments; (2) membranes; (3) bones; (4) hormone; (5) night; (6) Toxicity; (7) preformed; (8) supplements; (9) carotene; (10) absorption; (11) bone; (12) kidneys; (13) sun; (14) skin; (15) osteomalacia; (16) antioxidant; (17) pollutants; (18) hemolysis; (19) clotting; (20) bacteria; (21) newborns; (22) antibiotics.

Chapter study questions from the text—(1) Vitamins A, D, E, and K. Found in the fat and oily parts of foods; stored primarily in the liver and adipose tissue. Fat-soluble vitamins are stored longer; therefore, daily intake is less crucial and toxicity risk is greater than water-soluble vitamins. (2) Vitamin A is important in vision; maintenance of cornea, epithelial cells, mucous membranes, skin; bone and tooth growth; reproduction; hormone synthesis and regulation; immunity; cancer protection. Symptoms of deficiency include anemia, diarrhea, general discomfort, depression, frequent respiratory, digestive, bladder, vaginal and other infections, abnormal tooth and jaw alignment, night blindness, keratinization, corneal degeneration leading to blindness, rashes. (3) Compounds that can be converted into the active form of a vitamin. Carotenoids (the most active being beta-carotene) are available from plant foods; retinol compounds are available from animal foods. Examples of high vitamin A activity are: eye activity from light to dark, bone growth. (4) With sunlight, vitamin D can be made from cholesterol. Its chief function is to promote mineralization of bones. Richest sources are fortified milk, fortified margarine, eggs, liver, fatty fish. (5) Vitamin E protects other substances from oxidation by being oxidized itself; it protects the lipids and other vulnerable components of the cell and its membranes from destruction. It is especially effective in preventing the oxidation of the polyunsaturated fatty acids. Deficiency symptoms include red blood cell breakage and anemia. (6) Synthesis of blood-clotting proteins and a blood protein that regulates blood calcium. Conditions which can lead to vitamin K deficiency are taking of antibiotics or sulfa drugs while consuming a diet low in vitamin K.

Short Answers—
1. found in fat part of foods; stored primarily in liver and adipose tissue; need an average intake; toxicity is of concern
2. A; D; E; K

Problem Solving—
1. 7,930 IU divided by 10 = 793 RE (10 IU = 1 RE)
   45,390 IU divided by 3.33 = 13,631 RE (3.33 IU = 1 RE)
2. 9,120 RE divided by 3 oz = 3,040 RE/oz
   3,040 RE divided by 1,000 RE X 100 = 304%
   (U.S. RDA = 1,000 RE)

Sample Test Questions—

1.	a (p. 367)	6.	c (p. 371)	11.	c (p. 375)	16.	a (p. 381)
2.	b (p. 367)	7.	c (p. 372)	12.	b (p. 376)	17.	b (p. 381)
3.	a (p. 369)	8.	a (p. 372)	13.	d (p. 376)	18.	a (p. 381)
4.	e (p. 370)	9.	b (p. 373)	14.	b (p. 380)	19.	b (p. 382)
5.	a (p. 370)	10.	d (p. 375)	15.	d (p. 381)	20.	a (p. 382)

Crossword Puzzle—

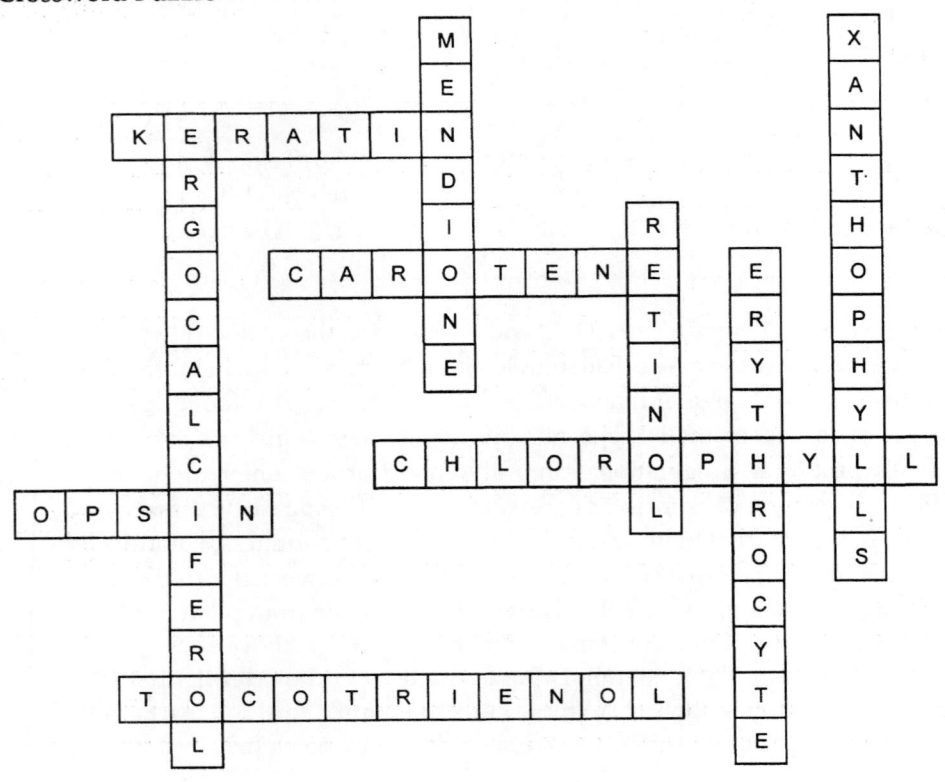

Chapter 12
Water and the Major Minerals

Chapter Outline

I. Water and the Body Fluids
    A. Water Balance and Recommended Intakes

    B. Blood Volume and Blood Pressure

    C. Fluid and Electrolyte Balance

    D. Fluid and Electrolyte Imbalance

    E. Acid-Base Balance

II. The Minerals—An Overview

III. Sodium

IV. Chloride

V. Potassium

_____
_____
_____

VI. Calcium
    A. Calcium Roles in the Body

_____
_____
_____

    B. Calcium Recommendations and Sources

_____
_____
_____

    C. Calcium Deficiency

_____
_____
_____

VII. Phosphorus

_____
_____
_____

VIII. Magnesium

_____
_____
_____

IX. Sulfur

_____
_____
_____

Highlight: Osteoporosis and Calcium

_____
_____
_____
_____

SUMMING UP

Water in the body participates in many chemical (1)_____, and serves as a solvent,

(2)_____ medium, and lubricant. It makes up (3)_____ percent of the body's weight. The principal (4)_____ in body fluids are sodium, chloride, phosphorus, and (5)_____; each is maintained at a constant concentration by means of renal excretion. The electrolytes are involved in maintaining the water and acid-base (6)_____. (7)_____ and electrolyte imbalances are medical emergencies. Sodium is abundant in the diet, as part of (8)_____. Deficiencies are rare except in (9)_____.

About 99 percent of the body's calcium is a structural component of the (10)_____ and teeth. The 1 percent of calcium found in body fluids helps maintain cell (11)_____ integrity, intercellular cohesion, transport of substances into and out of cells, and transmission of (12)_____ impulses. It is also essential for blood (13)_____ and acts as a cofactor in some enzyme systems. Calcium concentration in the blood is held constant.

Calcium deficiency may be caused directly by inadequate calcium intakes over a prolonged period or indirectly by vitamin (14)_____ deficiency. The diseases that result are rickets, osteomalacia, and (15)_____. A substantial portion of the recommended calcium intake is easily supplied by 2 cups of (16)_____ or equivalent dairy products such as cheese; (17)_____ soy milk is an alternative in the case of milk (18)_____ or lactose (19)_____.

Phosphorus, another major mineral, is so abundant in foods that (20)_____ are unlikely. It participates with calcium in forming the (21)_____ of bone and therefore is found in large quantities in the body. Magnesium plays a role in the synthesis of body (22)_____ and so is important to all body functions. It is found lacking in human beings in conditions that aggravate dietary protein deficiency, such as kwashiorkor and (23)_____. A deficiency of magnesium causes (24)_____. Sulfur, like phosphorus, is a major mineral constituent of body (25)_____. It is (26)_____ in the diet, and deficiencies are unknown.

Chapter Glossary

ADH (antidiuretic hormone): a hormone released by the pituitary gland in response to highly concentrated blood. The kidneys respond by reabsorbing water, thus preventing water loss.
adrenal glands: glands adjacent to, and just above, each kidney.
aldosterone: a hormone secreted by the adrenal glands that stimulates the reabsorption of sodium by the kidneys; aldosterone also regulates chloride and potassium concentrations.
angiotensin: a hormone involved in blood pressure regulation. Its precursor protein is called *angiotensinogen*.

anions: negatively charged ions.
artesian water: water drawn from a well that taps and confined aquifer in which the water is under pressure.
bicarbonate: a compound with the formula HCO_3 that results from the dissociation of carbonic acids; of particular importance in maintaining the body's acid-base balance.
binders: chemical compounds in foods that combine with nutrients (especially minerals) to form complexes the body cannot absorb.
bottled water: drinking water sold in bottles.
calcitonin: a hormone from the thyroid gland that regulates blood calcium by lowering it when levels rise too high.
calcium: the most abundant mineral in the body, found primarily in the body's bones and teeth.
calcium rigor: hardness or stiffness of the muscles caused by high blood calcium concentrations.
calcium-binding protein: a protein in the intestinal cells, made with the help of vitamin D, that facilitates calcium absorption.
calcium tetany: intermittent spasm of the extremities due to nervous and muscular excitability caused by low blood calcium concentrations.
calmodulin: an inactive protein that becomes active when bound to calcium; once activated, it becomes a messenger that tells other proteins what to do.
carbonic acid: a compound with the formula H_2CO_3 that results from the combination of carbon dioxide (CO_2) and water (H_2O), of particular importance in the body's buffer system.
cations: positively charged ions.
chloride: the major anion in the extracellular fluids of the body; chloride is the ionic forms of chlorine, Cl^-.
dehydration: the condition in which body water output exceeds water input.
dissociates: physically separates.
distilled water: water that has been vaporized and recondensed, leaving it free of dissolved minerals.
electrolyte solutions: solutions that can conduct electricity.
electrolytes: salts that dissolve in water and dissociate into charges particles called ions.
extracellular fluid: fluid outside the cells. Extracellular fluid includes interstitial fluids and plasma.
filtered water: water treated by filtration, usually through active carbon filters that reduce the lead in tap water, or by reverse osmosis units that force pressurized water across a membrane removing lead, arsenic, and some microorganisms from tap water.
fluid and electrolyte balance: maintenance of the proper amounts and kinds of fluid and minerals in each compartment of the body fluids.
hard water: water with a high calcium and magnesium content.
hydroxyapatite: crystals made of calcium and phosphorus.
hypothalamus: a brain center that controls activities such as maintenance of water balance and regulation of body temperature.
interstitial fluid: fluid between the cells, usually high in sodium and chloride. Interstitial fluid is a large component of *extracellular fluid* (fluid outside the cells).
intracellular fluid: fluid within the cells, usually high in potassium and phosphate; intracellular fluid accounts for approximately 2/3 of the body's water.
ions: atoms or molecules that have gained or lost electrons and therefore have electrical charges.
magnesium: a cation within the body's cells, active in many enzyme systems.
major minerals: essential mineral nutrients found in the human body in amounts larger than 5 grams; sometimes called macrominerals.
milliequivalents (mEq): the concentration of electrolytes in a volume of solution.

mineralization: the process in which calcium, phosphorus, and other minerals crystallize on the collagen matrix of a growing bone, hardening the bone.
minerals: inorganic elements; some minerals are essential nutrients required in small amounts.
mineral water: water from a spring or well that typically contains 250 to 500 parts per million (ppm) or minerals.
natural water: water obtained from a spring or well that is certified to be safe and sanitary.
obligatory water excretion: the amount of water the body has to excrete each day to dispose of its wastes.
oral rehydration therapy (ORT): a simple solution of sugar, salt, and water taken by mouth to treat dehydration.
osmosis: the movement of water across a membrane toward the side where the solutes are more concentrated.
osmotic pressure: the pressure that develops when two solutions of different concentrations are separated by a membrane that permits water, but not some of the solutes, to cross.
osteoporosis: a disease in which the bones become porous and fragile due to a loss of minerals; also called *adult bone loss*.
parathormone: a hormone from the parathyroid glands that regulates blood calcium by raising it when levels fall too low; also known as parathyroid hormone.
peak bone mass: the highest attainable bone density for an individual, developed during the first three decades of life.
pH: a measure of the concentration of H^+ ions.
phosphorus: a major mineral found mostly in the body's bones and teeth.
polar: a neutral molecule that has opposite charges spatially separated within the molecule.
potassium: the principal cation within the body's cells, critical to the maintenance of fluid balance, nerve transmissions, and muscle contractions.
public water: water from a municipal or county water system that has been treated and disinfected.
purified water: water that has been treated by distillation or other physical or chemical processes that remove dissolved solids.
renin: an enzyme from the kidneys that activates angiotensin.
salts: compounds composed of a positive ion other than H^+ and a negative ion other than OH^-.
salt sensitivity: a characteristic of individuals who respond to a high salt intake with an increase in blood pressure.
selectively permeable: cell membrane that permits some, but not all, substances to pass freely.
sodium: the principal cation in the extracellular fluids of the body, critical to the maintenance of fluid balance, nerve transmissions, and muscle contractions.
sodium-potassium pump: the pump activity that exchanges sodium for potassium across the cell membrane.
soft water: water with a high sodium or potassium content.
solutes: the substances that are dissolved in a solution.
spring water: water originating from an underground spring or well.
sulfur: a mineral present in the body as part of some amino acids.
thirst: a conscious desire to drink.
trace minerals: essential mineral nutrients found in the human body in amounts less than 5 grams.
vasoconstrictor: a substance that constricts or narrows the blood vessels.
water balance: the balance between water intake and output (losses).
water intoxication: the rare condition in which body water contents are too high.
well water: water drawn from ground water by tapping into an aquifer.

ASSIGNMENTS

Complete this crossword puzzle by Mary A. Wyandt, Ph.D., CHES.

Across:	Down:
2. water that has been treated by distillation or other physical or chemical processes	1. water from a spring or well that typically contains 250-500 ppm of minerals
7. water that has been vaporized and recondensed, leaving it free of dissolved minerals	3. water drawn from ground water by tapping in an aquifer
8. water originating from an underground spring or well	4. water treated by filtration, usually through active carbon filters that reduce lead in tap water
9. water from a municipal or county water system	5. water with a high calcium and magnesium content
10. water obtained from a spring or well that is certified to be safe and sanitary	6. water with a high sodium or potassium content

Answer these chapter study questions from the text:

1. List the roles of water in the body.

2. List the sources of water intake and routes of water excretion.

3. What is ADH? Where does it exert its action? What is aldosterone? How does it work?

4. How does the body use electrolytes to regulate fluid balance?

5. What do the terms *major* and *trace* mean when describing the minerals in the body?

6. Describe some characteristics of minerals that distinguish them from vitamins.

7. What is the major function of sodium in the body? Describe how the kidneys regulate blood sodium. Is a dietary deficiency of sodium likely? Why or why not?

8. List calcium's roles in the body. How does the body keep blood calcium constant regardless of intake?

9. Name significant food sources of calcium. What are the consequences of inadequate intakes?

10. List the roles of phosphorus in the body. Discuss the relationships between calcium and phosphorus. Is a dietary deficiency of phosphorus likely? Why or why not?

11. State the major functions of chloride, potassium, magnesium, and sulfur in the body. Are deficiencies of these nutrients likely to occur in your own diet? Why or why not?

Complete these short answer questions:

1. Name the major minerals.

    a.   e.

    b.   f.

    c.   g.

    d.

2. Three sources of water for the body are:

    a.

    b.

    c.

3. Four routes of water excretion are:

   a.

   b.

   c.

   d.

Solve these problems:

1. The soup recipe calls for 2 tsp of salt. How many grams of salt is that? How many grams of sodium are in that much salt?

2. If a cup of milk provides 300 mg of calcium and you are trying to consume 800 mg of calcium per day, how many cups of milk do you need to drink (assuming this is your only source of calcium)?

SAMPLE TEST QUESTIONS

1. The percentage of water in the body is about:

   a.   20%   c.   60%
   b.   40%   d.   90%

2. Water is involved in all of the following *except*:

   a.   regulation of body temperature.
   b.   conversion of lipids to amino acids.
   c.   shock absorber in the spinal cord.
   d.   lubrication of joints.

3. People must have water in their diets because the:

   a.   kidneys exert no control over the amount of water excreted in the urine.
   b.   fluid lost from the body must be replaced.
   c.   kidneys must excrete a minimum amount of water in the urine to rid the body of wastes.
   d.   a and b.
   e.   b and c.

4. Which of the following substances is an electrolyte?

    a. water
    b. sodium
    c. a fatty acid
    d. glucose
    e. carbon

5. A cell membrane is an example of:

    a. a selectively permeable membrane.
    b. a milliequivalent.
    c. an anion.
    d. an electrolyte.

6. The amount and type of electrolytes in the body are largely regulated by:

    a. the liver.
    b. the kidneys.
    c. the intestines.
    d. the spleen.
    e. the pancreas.

7. Aldosterone secretion stimulates:

    a. sodium retention.
    b. sodium excretion.
    c. water excretion.
    d. a and c.
    e. b and c.

8. Buffers:

    a. move water into a concentrated solution.
    b. transport oxygen in the blood.
    c. maintain acid-base balance in the body.
    d. maintain blood pressure in the body.

9. Foods that have the highest sodium content include:

    a. fresh vegetables.
    b. foods that taste salty.
    c. meat products.
    d. processed foods.

10. The major anion of extracellular fluid is:

    a. sodium.
    b. calcium.
    c. potassium.
    d. sulfur.
    e. chloride.

11. The most reliable food source of chloride is:

    a. meats and whole-grain cereals.
    b. salts.
    c. dark green vegetables.
    d. public water.
    e. milk and milk products.

12. The richest food sources of potassium are:

    a. processed foods.
    b. ready-to-eat cereals.
    c. fresh foods of all kinds.
    d. cured meats.

13. Blood calcium levels can be increased by:

    1. better absorption of calcium in the intestines.
    2. retention of calcium by the kidneys.
    3. release of calcium from the bones.
    4. manufacture of calcium from albumin.
    5. release of calcium from the teeth.

    a. 1, 2
    b. 1, 2, 3
    c. 1, 2, 3, 5
    d. 3, 4

14. Factors that impair calcium balance include:

    a. vitamin D.
    b. high-fiber diet.
    c. phytates.
    d. a and b.
    e. b and c.

15. An inadequate intake of calcium for many years can lead to:

    a. osteoporosis.
    b. impaired energy-nutrient metabolism.
    c. pernicious anemia.
    d. bleeding gums and ease in bruising.
    e. peptic ulcers.

16. A class of foods that is rich in calcium and phosphorus is:

    a. citrus fruits.
    b. milk and milk products.
    c. potato.
    d. carrots.
    e. liver and other organ meats.

17. Positive calcium balance is favored by:

    a. high-fat intake.
    b. fiber.
    c. vitamin D.
    d. a and b

18. Phosphorus deficiencies are:

    a. rare.
    b. common.
    c. irreversible.
    d. a and b.
    e. b and c.

19. Magnesium:

    a. is directly necessary for protein synthesis in cells.
    b. protects bone structures against degeneration.
    c. is the body's principal intracellular electrolyte.
    d. is necessary for wound healing.
    e. helps maintain gastric acidity.

20. Sulfur:

    a. is found in the adipose tissue.
    b. is commonly deficient.
    c. has an RDA of 400 mg.
    d. is present in all proteins.

Answers

Summing Up—(1) reactions; (2) transportation; (3) 55 to 60; (4) electrolytes; (5) potassium; (6) balance; (7) Fluid; (8) salt; (9) dehydration; (10) bones; (11) membrane; (12) nerve; (13) clotting; (14) D; (15) osteoporosis; (16) milk; (17) fortifies; (18) allergy; (19) intolerance; (20) deficiencies; (21) crystals; (22) proteins; (23) alcoholism; (24) tetany; (25) tissues; (26) abundant.

Crossword Puzzle—

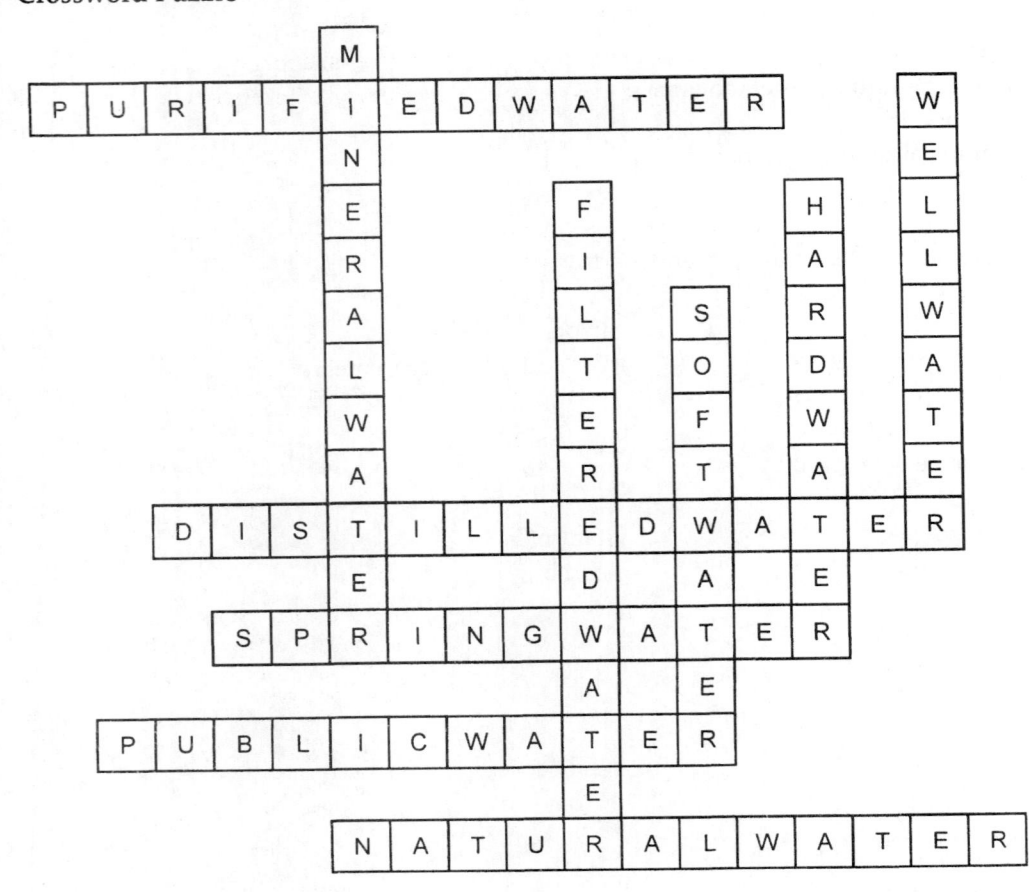

- PURIFIED WATER
- DISTILLED WATER
- SPRING WATER
- PUBLIC WATER
- NATURAL WATER
- MINERAL WATER
- FILTERED WATER
- SOFT WATER
- HARD WATER
- WELL WATER

Chapter study questions from the text—(1) Water carries nutrients and waste products throughout the body; helps to form the structure of macromolecules; actively participates in chemical reactions; fills the cells and the spaces between them; serves as the solvent for minerals, vitamins, amino acids, glucose, and many other small molecules; acts as lubricant around joints; serves as shock absorber inside the eyes, spinal cord, and in pregnancy, the amniotic sac; aids in the body's temperature regulation. (2) Sources of intake: liquids, foods, metabolic water. Routes of excretion: kidneys, skin, lungs, feces. (3) ADH is a hormone released by the pituitary gland in response to highly concentrated blood. It exerts its action on the kidneys, and they respond by reabsorbing water, thus preventing water loss. Aldosterone is a hormone secreted by the adrenal glands that stimulates the reabsorption of sodium by the kidneys. (4) The body uses electrolytes to control the movement of water inside cells and between cells. (5) Major minerals are needed in the largest amounts in the body; trace minerals are needed in small amounts in the body. Both are equally important. (6) Minerals can become charged particles and can form compounds; minerals retain their chemical identity. (7) To maintain fluid and electrolyte balance. Kidneys retain or excrete sodium and water to regulate blood pressure. A dietary deficiency of sodium is not likely because of sodium's abundance in food products. (8) Bone structure, cell membrane integrity, transport of ions, muscle action, nerve impulses, regulates blood vessel wall muscle tone, helps regulate blood pressure, aids blood clotting, acts as cofactor for enzymes. When levels fall, intestinal absorption increases, bone withdrawal increases, and kidney excretion diminishes; these processes are regulated by a system of hormones and vitamin D. (9) Calcium is found predominantly in milk and milk products. Inadequate intakes limit the bones' ability to achieve optimal mass and density and increase risk of osteoporosis and associated fractures. (10) Bone structure, part of DNA and RNA, activates enzymes, part of ATP, part of lipid structure and cell membranes. They combine as calcium phosphate in the crystals of bone and teeth, providing strength and rigidity. A deficiency is unlikely because phosphorus is abundant in animal tissues, eggs, and milk. (11) Chloride: maintains pH balance, allows nerve transmission and muscle contraction, catalyst in metabolism; magnesium: bone structure, protein synthesis, energy metabolism, muscle relaxation; sulfur: protein structure. Deficiency is unlikely in these minerals because they naturally occur in many foods.

Short Answers—
1. calcium; phosphorus; potassium; sulfur; sodium; chloride; magnesium
2. liquids; foods; metabolic water
3. kidneys; lungs; feces; skin

Problem Solving—
1. 2 tsp X 5 g/tsp = 10 g salt

    10 g salt X 2 g sodium/5 g salt = 4 g sodium

2. 800 mg divided by 300 mg/c = 2.7 c (rounded)

Sample Test Questions—
1.	c (p. 395)	6.	b (p. 403)	11.	b (p. 410)	16.	b (p. 419)
2.	b (p. 396)	7.	a (p. 404)	12.	c (p. 411)	17.	e (p. 415)
3.	e (p. 397)	8.	c (p. 404)	13.	b (p. 413)	18.	a (p. 419)
4.	b (p. 401)	9.	d (p. 408)	14.	e (p. 415)	19.	a (p. 420)
5.	a (p. 402)	10.	e (p. 410)	15.	a (p. 417)	20.	d (p. 422)

CHAPTER 13
THE TRACE MINERALS

CHAPTER OUTLINE

I. The Trace Minerals—An Overview

II. Iron
   A. Iron Roles in the Body

   B. Iron Absorption and Metabolism

   C. Iron Deficiency

   D. Iron Toxicity

   E. Iron Recommendations and Sources

   F. Iron Contamination and Supplementation

III. Zinc
   A. Zinc Roles in the Body

B. Zinc Absorption and Metabolism

C. Zinc Deficiency

D. Zinc Toxicity

E. Zinc Recommendations and Intakes

F. Zinc Supplementation

IV. Iodine

V. Selenium

VI. Copper

VII. Manganese

VIII. Fluoride

IX. Chromium

X. Molybdenum

XI. Other Trace Minerals

XII. Contaminant Minerals

XIII. Closing Thoughts on the Nutrients

Highlight: Phytochemicals and Functional Foods

SUMMING UP

Iron is found principally in the red blood cells, where it comprises part of the (1)_____-carrier protein hemoglobin. When red blood cells die and are dismantled in the (2)_____, the iron is retrieved and transported by iron-carrier proteins back to bone (3)_____, where new red blood cells are synthesized. There is no route of excretion for iron; losses are small except when (4)_____ is lost, as in (5)_____ or hemorrhage. Thus women's needs for iron are ordinarily (6)_____ than men's.

Iron-deficiency (7)_____, one of the world's most widespread malnutrition problems, is most common in women and (8)_____. Furthermore, (9)_____ iron deficiency is a vast world malnutrition problem, limiting people's working ability. Food sources of iron for women must be chosen carefully if even two-thirds of the recommended intake is to be met within a (10)_____ allowance that is not excessive. The (11)_____ of foods somewhat improves women's iron intakes, but may produce iron (12)_____ in men. Addition of an iron (13)_____ to the diet may be advisable for some women.

(14)_____, fish, and poultry are superior sources of iron. (15)_____-grain or enriched breads and cereals, legumes, dark greens, and some fruits are other good sources of iron. Foods in the (16)_____ group are notable for their lack of iron. Iron absorption is enhanced by vitamin (17)_____, other organic acids, sugars, and the (18)_____ factor; it is reduced by phytates, fibers, soy, legumes, (19)_____, and coffee. Zinc appears in every body tissue and supports several physiological functions, including normal growth and (20)_____ development. As with iron, a homeostatic mechanism may regulate the amount of zinc (21)_____ by the body. (22)_____ of zinc have been observed in some children in the United States and Canada as well as in the Middle East. The richest food sources of zinc include (23)_____, meat, and liver. Milk, eggs, and whole-grain products are also good sources. When estimating the amount of zinc in a diet, one must take into account all of the factors which affect its (24)_____.

Iodide forms part of the (25)_____ hormones; deficiency may cause simple (26)_____, slowed metabolism, and cretinism. The use of iodized (27)_____ protects against deficiency. Copper is important for red blood cell formation, (28)_____ synthesis, and central nervous system function.

Manganese aids many body (29)_____, but its safe range is narrow, with toxicity causing a severe (30)_____-disease syndrome in human beings. The (31)_____ ion combines with calcium and phosphorus to stabilize the (32)_____ structure of bones and teeth. In communities where the water contains fluoride, dental (33)_____ are less prevalent than in communities where the water supply is low in fluoride. Chromium, as part of the (34)_____ tolerance factor, works with (35)_____ in promoting glucose uptake into cells and normal carbohydrate metabolism. Chromium deficiency is believed to be responsible for some cases of adult-onset (36)_____ and to cause growth failure in children with protein-energy malnutrition.

Selenium acts as cofactor for an (37)_____ enzyme. A severe deficiency can cause (38)_____ failure. Selenium-poor soil may correlate with certain kinds of (39)_____.

Molybdenum functions as a part of several (40)_____ systems. Other (41)_____ minerals with known physiological roles include nickel, silicon, tin, vanadium, and cobalt.

CHAPTER GLOSSARY

chelate: a substance that can grasp the positive ions of a metal.
cofactor: a mineral element that, like a coenzyme, works with an enzyme to facilitate a chemical reaction.
contamination iron: iron found in foods as the result of contamination by inorganic iron salts from iron cookware, iron-containing soils, and the like.
cretinism: a congenital disease characterized by mental and physical retardation and commonly caused by iodine deficiency during pregnancy.
enteropancreatic circulation: the circulatory route from the pancreas to the intestine and back to the pancreas.
erythrocyte protoporphyrin: a precursor to hemoglobin.
ferric iron: the oxidized ionic state of iron; Fe+++.
ferritin: an iron-storage protein.
ferrous iron: the reduced ionic state of iron; Fe++.
fluorapatite: the stabilized form of bone and tooth crystal, in which fluoride has replaced the hydroxyl groups of hydroxyapatite.
fluorosis: discoloration and pitting of tooth enamel caused by excess fluoride during tooth development.
galvanized: a term referring to metals that have been treated with a zinc-containing coating to prevent rust.
geophagia: a craving for and eating of clay.
glucose tolerance factor (GTF): a small organic compound that enhances insulin's action.
goiter: an enlargement of the thyroid gland due to an iodine deficiency, malfunction of the gland, or over-consumption of a goitrogen.
goitrogen: a thyroid antagonist found in food; causes *toxic goiter*.
heavy metals: any number of mineral ions such as mercury and lead, so called because they are of relatively high atomic weight; many heavy metals are poisonous.
hematocrit: measurement of the volume of the red blood cells packed by centrifuge in a given volume of blood.
heme: the iron-holding part of the hemoglobin and myoglobin proteins.
hemochromatosis: a hereditary defect in iron metabolism characterized by deposits of iron-containing pigment in many tissues, with tissue damage.
hemoglobin: the oxygen-carrying protein of the red blood cells that transport oxygen from the lungs to tissues throughout the body.
hemosiderin: an iron storage protein primarily made in times of iron overload.
hemosiderosis: a condition characterized by the deposition of hemosiderin in the liver and other tissues.
iron deficiency: the state of having depleted iron stores.
iron-deficiency anemia: a severe depletion of iron stores that result in low hemoglobin and small, pale, red blood cells.

iron overload: toxicity from excess iron.
Keshan disease: the heart disease associated with selenium deficiency.
metalloenzymes: enzymes that contain one or more minerals as part of their structures.
metallothionein: a sulfur-rich protein that avidly binds with metals such as zinc.
MFP factor: a factor associated with the digestion of meat, fish, and poultry that enhances iron absorption.
microcytic hypochromic anemia: anemia characterized by small, pale cells.
molybdenum: a trace element.
mucosa: a mucous membrane such as the one that lines the digestive tract.
mucosal ferritin: a protein in the mucosal cells that holds iron in the cells.
mucosal transferrin: a protein in the mucosal cells that passes iron on to blood transferrin.
myoglobin: the oxygen-holding protein of the muscle cells.
pagophagia: a craving for ice.
pica: a craving for nonfood substances.
RBC count: measurement of the number of red blood cells.
selenium: a trace element.
simple goiter: goiter caused by iodine deficiency.
thyroxine: released by the thyroid gland to target tissues.
trace minerals: essential mineral nutrients found in the human body in amounts smaller than 5 grams; sometimes called microminerals.
transferrin: the iron transport protein.

Assignments

Answer these chapter study questions from the text:

1. Distinguish between heme and nonheme iron. Discuss the factors that enhance iron absorption.

2. Distinguish between iron deficiency and iron-deficiency anemia. What are the symptoms of iron-deficiency anemia?

3. What causes iron overload? What are its symptoms?

4. Describe the similarities and differences in the absorption and regulation of iron and zinc.

5. Discuss possible reasons for a low intake of zinc. What factors affect the bioavailability of zinc?

6. Describe the principal functions of iodide, selenium, copper, manganese, fluoride, chromium and molybdenum in the body.

7. What public health measure has been used in preventing simple goiter? What measure has been recommended for protection against tooth decay?

8. Discuss the importance of balanced and varied diets in obtaining the essential minerals and avoiding toxicities.

9. Describe some of the ways trace minerals interact with each other and with other nutrients.

Complete these short answer questions:

1. Name the trace minerals.

   a.	f.	k.
   b.	g.	l.
   c.	h.	m.
   d.	i.	n.
   e.	j.

2. Iron's two ionic states are:

   a.	b.

Complete this crossword puzzle by Mary A. Wyandt, Ph.D., CHES.

	Across:		Down:
3.	a condition characterized by the deposition of hemosiderin in the liver and other tissues	1.	a congenital disease characterized by mental and physical retardation and commonly caused by iodine deficiency during pregnancy
5.	a trace element (symbol: Mo)	2.	the iron-holding part of the hemoglobin and myoglobin proteins
7.	the oxygen-carrying protein of the red blood cells that transport oxygen from the lungs to tissues throughout the body	4.	a trace element (symbol: Se)
8.	measurement of the volume of the red blood cells packed by centrifuge in a given volume of blood	5.	the oxygen-holding protein of the muscle cells
9.	the stabilized form of bone and tooth crystal	6.	an enlargement of the thyroid gland due to iodine deficiency

SAMPLE TEST QUESTIONS

1. Iron is important in the body because it is:

   a. needed for blood clotting.
   b. an integral part of bones and teeth.
   c. a constituent of hemoglobin.
   d. an antioxidant.

2. Which of the following iron-containing compounds transports iron from the intestine to the liver and bone marrow?

   a. ferritin
   b. myoglobin
   c. transferrin
   d. hemoglobin

3. The most common symptom of an iron deficiency is:

   a. night blindness.
   b. abnormal blood clotting.
   c. anemia.
   d. a skin rash.
   e. hemophilia.

4. Geophagia is:

   a. clay-eating.
   b. a mineral deficiency associated with a certain area of the country.
   c. a localized skin rash.
   d. a cancer-causing additive.

5. The hemoglobin level in the blood is used to assess a person's _____ status.

   a. copper
   b. folate
   c. vitamin B$_{12}$
   d. iron
   e. zinc

6. Which food group supplies most of the iron in the American diet?

   a. meat group
   b. milk group
   c. vegetable-fruit group
   d. bread-cereal group

7. A very poor source of iron is:

   a. legumes.
   b. dried fruits.
   c. whole grain breads.
   d. milk.

8. An important function of zinc is to serve as a(n):

   a. enzyme.
   b. cofactor.
   c. protein carrier.
   d. oxygen carrier.

9. Zinc is involved in the circulatory route from the pancreas to the intestine and back to the pancreas. This is called:

    a. phytate enhancement.
    b. oxalate-binding cofactor.
    c. ligand-operating absorption.
    d. cofactor assistance.
    e. enteropancreatic circulation.

10. Zinc deficiency symptoms include all but one of the following:

    a. severe growth retardation.
    b. arrested sexual maturation.
    c. decreased taste sensitivity.
    d. pernicious anemia.

11. Among the following, the best sources of available zinc are:

    a. shellfish, meats, and liver.
    b. breads, cereals, and grains.
    c. fruits and vegetables.
    d. milk products.

12. Consumption of which of the following would best help to insure normal iodide intake?

    a. liver
    b. irradiated milk
    c. salt-water fish
    d. fortified margarine

13. Cretinism is caused by a deficiency of:

    a. copper.
    b. iron.
    c. fluoride.
    d. zinc.
    e. iodide.

14. One of copper's roles is to:

    a. function as an antioxidant.
    b. manufacture collagen.
    c. assist in oxidation of ferrous iron to ferric iron.
    d. a and b.
    e. a, b, and c.

15. Manganese deficiencies are common in children.

    a. True
    b. False

16. Fluoride is necessary nutritionally for:

    a. hardness of the bones and teeth.
    b. production of the thyroid hormone.
    c. prevention of anemia.
    d. the metabolism of glucose.
    e. a and b.

17. Excess fluoride can cause:

    a. deposition of calcium in soft tissues.
    b. a drastic increase in tooth decay.
    c. fluorosis.
    d. osteomalacia.

18. Selenium is involved in:

    a. hormone production.
    b. protein synthesis.
    c. antioxidant activities.
    d. glycogen breakdown.

19. Contaminant minerals impair the body's growth and work capacity. They:

    a. enter the food supply by way of soil, water and air pollution.
    b. include lead and mercury.
    c. are food additives.
    d. a and b.
    e. a, b and c.

20. Cobalt seems to be important in nutrition as part of:

    a. vitamin A.
    b. glucose tolerance factor.
    c. vitamin B_{12}.
    d. vitamin B_6

Answers

Summing Up—(1) oxygen; (2) liver; (3) marrow; (4) blood; (5) menstruation; (6) greater; (7) anemia; (8) children; (9) marginal; (10) kcalorie; (11) enrichment; (12) overload; (13) supplement; (14) Meats; (15) Whole; (16) milk; (17) C; (18) MFP; (19) tea; (20) sexual; (21) absorbed; (22) Deficiencies; (23) shellfish; (24) availability; (25) thyroid; (26) goiter; (27) salt; (28) collagen; (29) enzymes; (30) brain; (31) fluoride; (32) crystalline; (33) caries; (34) glucose; (35) insulin; (36) diabetes; (37) antioxidant; (38) heart; (39) cancer; (40) enzyme; (41) trace.

Chapter study questions from the text—(1) Heme iron is contained in the organic molecule heme and is better absorbed than non-heme iron. Vitamin C, organic acids, sugars, and MFP factor enhance iron absorption. (2) Iron deficiency is the state of being without iron stores; iron-deficiency anemia is small and pale blood cells resulting from an iron deficiency causing pallor, fatigue, weakness, headaches, and apathy. (3) Iron overload is usually caused by a gene that enhances iron absorption. Symptoms: fatigue, headache, irritability, lowered work performance, anemia. (4) Similarities exist in absorption and regulation: absorption of both is partially determined by a person's status; other dietary factors can inhibit or facilitate absorption. Differences exist in that the body receives zinc from both ingested food and zinc-rich pancreatic secretions; the storage methods for each vary. (5) Diets very low in animal protein may be inadequate in zinc. Phytates and fiber. (6) Iodide: part of thyroid hormones; selenium: part of enzyme that acts as an antioxidant; copper: part of enzymes, catalyst in hemoglobin formation, collagen synthesis and wound healing, maintains nerve fiber sheaths; manganese: acts as a cofactor for many enzymes that facilitate the metabolism of carbohydrate, lipids, and amino acids; fluoride: part of crystal structure of bones and teeth; chromium: part of glucose tolerance factor; molybdenum: acts as a working part of several metalloenzymes. (7) Iodization of salt. Fluoridation of water. (8) Such a diet prevents an excess of one trace mineral from causing a deficiency of another, or of a deficiency which may cause a toxic reaction of another, provides factors that promote trace mineral absorption, and includes food sources that contain all the trace minerals. (9) Fiber and phytates bind zinc, limiting its bioavailability; large doses of iron inhibit zinc absorption; large doses of zinc inhibit iron and copper absorption.

Short Answers--
1. iron; zinc; iodide; copper; manganese; fluoride; chromium; selenium; molybdenum; nickel; silicon; tin; cobalt; arsenic
2. ferrous (reduced, Fe++); ferric (oxidized, Fe+++)

Crossword Puzzle—

```
                    H                               C
                    E                               R
                    M                               E
      H   E   M   O   S   I   D   E   R   O   S     T
                    E                               I   S
                    L                               N
M   O   L   Y   B   D   E   N   U   M               I
Y                   N                               S
O                   I                               M
G                   U           G
L           H   E   M   O   G   L   O   B   I   N
O                               I
B               H   E   M   A   T   O   C   R   I   T
I                               E
N               F   L   U   O   R   A   P   A   T   I   T   E
```

Sample Test Questions—

1. c (p. 439)
2. c (p. 440)
3. c (p. 442)
4. a (p. 443)
5. d (p. 442)
6. a (p. 445)
7. d (p. 445)
8. b (p. 447)
9. e (p. 448)
10. d (p. 449)
11. a (p. 449)
12. c (p. 452)
13. e (p. 451)
14. e (p. 454)
15. b (p. 455)
16. a (p. 455)
17. c (p. 456)
18. c (p. 453)
19. d (p. 458)
20. c (p. 458)

Chapter 14
Fitness: Physical Activity, Nutrients, and Body Adaptations

Chapter Outline

I. Fitness
    A. Benefits of Fitness

    B. Developing Fitness

    C. Cardiorespiratory Endurance

II. Energy Systems, Fuels, and Nutrients to Support Activity
    A. The Energy Systems of Physical Activity—ATP and CP

    B. Glucose Use during Physical Activity

    C. Fat Use during Physical Activity

    D. Protein Use during Physical Activity--and between Times

    E. Vitamins and Minerals to Support Activity

  F. Fluids and Electrolytes to Support Activity

_____
_____
_____
_____

  G. Poor Beverage Choices: Caffeine and Alcohol

_____
_____
_____
_____

III. Diets for Physically Active People
  A. Choosing a Diet to Support Fitness

_____
_____
_____
_____

  B. Meals Before and After Competition

_____
_____
_____
_____

Highlight: Supplements as Ergogenic Aids

_____
_____
_____
_____

SUMMING UP

In the body, nutrition and (1)_____ go hand in hand. The working body demands that the energy-yielding nutrients provide (2)_____ for the increased metabolism of exercise. Lack of exercise can lead to cardiovascular disease, (3)_____, intestinal disorders, apathy, (4)_____, and accelerated (5)_____ losses. The four components of fitness are flexibility, (6)_____, muscle endurance, and (7)_____ endurance.

Exercises that improve cardiovascular endurance raise the heart rate for more than (8)_____ minutes and use most of the large muscle groups of the body. Exercises are classified as (9)_____ and anaerobic.

The first fuels of exercise are adenosine triphosphate (ATP) and (10)_____ (PC). Glucose, stored as (11)_____, is essential for performance. A high (12)_____ diet is most beneficial for athletic performance. Intensity and (13)_____ of exercise affect use of exercise fuels.

Vitamins and minerals assist in releasing energy from fuels and transporting (14)_____. An adequate diet will usually supply the necessary vitamins and minerals for athletes and (15)_____ are generally not recommended. Oversupplementation may actually (16)_____ athletic performance.

Dehydration is a concern for athletes. Full hydration is imperative for athletes during (17)_____ as well as during competition. Plain water or diluted juice is recommended. (18)_____ loading is used to trick the muscles into storing extra glycogen before competition. This technique has been altered to minimize undesirable side effects.

A diet that provides ample (19)_____ and consists of a variety of nutrient dense foods in quantities to meet (20)_____ needs will not only enhance athletic performance but overall health as well.

Chapter Glossary

amenorrhea: the absence of or cessation of menstruation.
athletic amenorrhea: cessation of menstruation associated with strenuous athletic training.
atrophy: of muscles, becoming smaller; a decrease in size because of disuse, undernutrition, or wasting diseases.
body composition: the proportions of muscle, bone, fat and other tissue that make up a person's total body weight.
carbohydrate loading: a regimen of moderate exercise followed by consuming a high-carbohydrate diet that enables muscles to store glycogen beyond their normal capacity; also called glycogen loading or glycogen supercompensation.
cardiac output: the volume of blood discharged by the heart each minute; determined by multiplying the stroke volume by the heart rate.
cardiorespiratory conditioning: improvements in heart and lung function and increased blood volume, brought about by aerobic training.
cardiorespiratory endurance: the ability to perform large-muscle, dynamic exercise of moderate-to-high intensity for prolonged periods.
conditioning: the physical effect of training; improved flexibility, strength, and endurance.
cool-down: five to ten minutes of light activity such as walking or stretching following a vigorous workout to return the body's core gradually to near-normal temperature.
CP creatine phosphate (also called phosphocreatine): a high-energy compound in muscle cells that acts as a reservoir of energy that can maintain a steady supply of ATP; CP provides the energy for short bursts of activity.

duration: length of time (for example, the time spent in each exercise session).

epinephrine: one of the stress hormones that is secreted whenever emergency action is called for; it readies body systems for fast action and mobilizes fuel to support that action.

exercise: planned, structured, and repetitive bodily movement that promotes or maintains physical fitness.

fitness: the characteristics that enable the body to perform physical activity; more broadly, the ability to meet routine physical demands with enough reserve energy to rise to a physical challenge; or the body's ability to withstand stress of all kinds.

flexibility: the capacity of the joints to move through a full range of motion; the ability to bend and recover without injury.

frequency: the number of occurrences per unit of time (for example, the number of exercise sessions per week).

glucose polymers: compounds that supply glucose, not as single molecules, but linked in chains somewhat like starch. The objective is to attract less water from the body into the digestive tract.

heat stroke: the dangerous accumulation of body heat with accompanying loss of body fluid.

hyperthermia: an above-normal body temperature.

hypertrophy: of muscles, growing larger; an increase in size in response to use.

hyponatremia: a decreased concentration of sodium in the blood.

hypothermia: a below-normal body temperature.

intensity: the degree of exertion while exercising (for example, the amount of weight lifted or the speed of running).

lactic acid: the anaerobic breakdown products of pyruvate.

mitochondria: the structures within a cell responsible for producing ATP.

moderate exercise: activity that can be sustained comfortably for 60 minutes or so.

muscle endurance: the ability of a muscle to contract repeatedly without becoming exhausted.

muscle fibers: muscle cells.

muscle strength: the ability of muscles to work against resistance.

physical activity: bodily movement produced by muscle contractions that substantially increase energy expenditure.

progressive overload principle: the training principle that a body system, in order to improve, must be worked at frequencies, durations, or intensities that gradually increase physical demands.

sedentary: physically inactive (literally, "sitting down a lot").

sports anemia: a transient condition of low hemoglobin in the blood, associated with early stages of sports training or other strenuous activity.

stroke volume: the amount of oxygenated blood the heart ejects toward the tissues at each beat.

training: practicing an activity regularly, which leads to conditioning (training is what you do; conditioning is what you get).

VO$_2$ max: the maximum rate of oxygen consumption by an individual at sea level.

warm-up: five to ten minutes of light activity, such as easy jogging or cycling, prior to a workout to prepare the body for more vigorous activity.

weight training: the use of free weights or weight machines to provide resistance for developing muscle strength and endurance. A person's own body weight may also be used to provide resistance as when a person does push-ups, pull-ups, or abdominal crunches.

Assignments

Answer these chapter study questions from the text:

1. Define fitness, and list its benefits.

2. Explain the overload principle.

3. Define cardiorespiratory endurance and list some of its benefits.

4. What types of exercise are aerobic? Which are anaerobic?

5. Describe the relationships among energy expenditure, type of activity, and oxygen use.

6. What factors influence the body's use of glucose during physical activity? How?

7. What factors influence the body's use of fat during physical activity? How?

8. What factors influence the body's use of protein during physical activity? How?

9. Why are some athletes likely to develop iron-deficiency anemia? Compare iron-deficiency anemia and sports anemia, explaining the differences.

10. Discuss the importance of hydration during training, and list recommendations to maintain fluid balance.

11. Describe the components of a healthy diet for athletic performance.

Complete these short answer questions:

1. The components of fitness are:

    a.   c.

    b.   d.

2. Physical activity is classified as:

    a.   b.

3. Cardiovascular conditioning is characterized by:

    a.

    b.

    c.

    d.

    e.

    f.

Complete this crossword puzzle by Mary A. Wyandt, Ph.D., CHES.

Across:	Down:
1. practicing an activity regularly, which leads to conditioning	2. the degree of exertion while exercising
6. length of time; for example, the time spent in each exercise session	3. of muscles, growing larger; an increase in size in response to use
7. physically inactive (literally, "sitting down a lot")	4. of muscles, becoming smaller; a decrease in size because of disuse, under-nutrition, or wasting disease
9. the capacity of the joints to move through a full range of motion; the ability to bend and recover without injury	5. the physical effect of training; improved flexibility, strength, and endurance
10. the ability of muscles to work against resistance	8. the characteristics that enable the body to perform physical activity; more broadly, the ability to meet routine physical demands with enough reserve energy to rise to a physical challenge; or the body's ability to withstand stress of all kinds

SAMPLE TEST QUESTIONS

1. Lack of exercise can lead to development of:

   a. cardiovascular disease.
   b. obesity.
   c. intestinal disorders.
   d. accelerated bone losses.
   e. all of the above.

2. The components of fitness are:

   a. flexibility.
   b. strength.
   c. muscle endurance.
   d. cardiorespiratory endurance.
   e. all of the above.

3. Muscle response to disuse or undernutrition is called:

   a. hypertrophy.
   b. atrophy.
   c. slow-twitch fibers.
   d. fast-twitch fibers.

4. The maximum rate of oxygen consumption is:

   a. aerobic.
   b. anaerobic.
   c. VO$_2$ max.
   d. cardiac output.

5. Lactic acid is a product of _____ metabolism.

   a. aerobic
   b. anaerobic

6. Factor(s) that influence glycogen use during physical activity include:

   a. amount of carbohydrate in the diet.
   b. intensity and duration of the activity.
   c. degree of training to perform the activity.
   d. all of the above
   e. a and b

7. During low or moderate intensity activity, muscle cells use predominantly _____ as fuel.

   a. glycogen
   b. fat
   c. protein
   d. water
   e. none of the above

8. Fat can be used for fuel by muscles if the work is aerobic.

   a. True
   b. False

9. The recommended protein intake for athletes is much higher than for non-athletes.

    a. True b. False

10. Athletes can safely add muscle tissue by:

    a. tripling their protein intake.
    b. taking hormones duplicating those of puberty.
    c. putting a demand on muscles making them work harder.
    d. depending on protein for muscle fuel and cutting down on carbohydrates.

11. Which factor(s) modify/ies the body's use of protein?

    a. amount of carbohydrate in the diet
    b. exercise intensity and duration
    c. the degree of training
    d. a and b
    e. a, b, and c

12. This element is important in transport of oxygen in blood and in muscle tissue and energy transformation reactions.

    a. vitamin C   d. calcium
    b. zinc   e. thiamin
    c. iron

13. An above-normal body temperature is

    a. hyperthermia.   c. hypothermia.
    b. heat stroke.

14. Alcohol is a diuretic and therefore it induces:

    a. energy-release.   c. improved performance.
    b. fluid losses.

15. The best choice for an athlete who needs to rehydrate is:

    a. concentrated juice. d. sweat replacers.
    b. cool water.   e. a or b
    c. salt tablets

16. This type of athlete is prone to iron deficiency:

    a. male weight lifter. c. female endurance athlete.
    b. male swimmer.   d. female sprinter.

17. A transient condition of low hemoglobin in the blood associated with early stages of strenuous athletic training is:

   a. sports anemia.
   b. menopause.
   c. stress menstruation.
   d. amenorrhea.

18. A healthy diet for athletes consists of:

   1. nutrient dense foods.
   2. vitamin and mineral supplements.
   3. ample fluids.
   4. salt tablets.
   5. adequate food to meet energy requirements.
   6. adequate amounts of vitamins and minerals.
   7. protein powders.

   a. 1, 2, 4, 7
   b. 1, 3, 4, 5
   c. 1, 3, 5, 6
   d. 2, 3, 4, 7
   e. 2, 4, 5, 6

19. Alcohol hinders athletic activity by:

   a. acting as a diuretic.
   b. stimulating the central nervous system.
   c. altering perceptions.
   d. a and b
   e. a and c

20. The recommended pregame meal includes:

   a. vitamin/mineral supplement.
   b. plenty of fluids.
   c. light, easily digestible foods.
   d. b and c
   e. a, b, and c

Answers

Summing Up—(1) exercise; (2) fuel; (3) obesity; (4) insomnia; (5) bone; (6) strength; (7) cardiovascular; (8) 20; (9) aerobic; (10) phosphocreatine; (11) glycogen; (12) carbohydrate; (13) duration; (14) oxygen; (15) supplements; (16) impair; (17) training; (18) glycogen; (19) fluid; (20) energy.

Chapter study questions from the text—(1) Fitness: the body's ability to meet physical demands. Benefits: more restful sleep, improved nutritional health, reduced fatness and increased lean body tissue, greater bone density, improved resistance to infectious diseases, improved circulation and lung function, reduced risk of some cancers, reduced risk of diabetes, reduced incidence and severity of anxiety and depression, improved self-image and self-confidence, long life and improved quality of life. (2) The training principle that a body system, in order to improve, must be worked at frequencies, durations, or intensities that gradually increase physical demands. (3) The ability to perform large-muscle dynamic exercise of moderate-to-high intensity for prolonged periods. Benefits include support of ongoing action of the heart and lungs. (4) Aerobic: requiring oxygen (swimming, cross-country skiing, rowing, fast walking, jogging, fast bicycling, soccer, hockey, basketball, water polo, lacrosse and rugby). Anaerobic: not requiring oxygen (golf—riding, not walking; ballet dancing; double tennis). (5) Slow-twitch muscle fibers are best suited to producing energy by aerobic processes for prolonged endurance exercise. Fast-twitch muscle fibers are best suited to producing energy by anaerobic processes for high intensity, short-duration work. (6) The body's use of glucose during physical activity depends partially on how much glycogen is in storage and this depends partly on the amount of carbohydrate eaten. Intensity of exercise influences the body's use of glucose—high intensity activities require more glycogen. Degree of training to perform the activity is a factor because the level of oxygen in the muscle influences the body's use of glucose. During oxygen debt, glucose is metabolized rapidly and pyruvate molecules accumulate in the muscle tissue. Duration of activity affects use of glucose during exercise—within the first 20 minutes of exercise, the body primarily uses glucose. (7) Duration of activity—when a moderate activity progresses past 20 minutes, the body uses more of its stored body fat for energy; intensity of activity—as the intensity increases, fat makes less and less of a contribution to the mixture of fuel used; degree of training—a well-trained body produces the adaptations that permit the body to draw heavily on fat for fuel. (8) Diet—people who consume diets rich in carbohydrate use less protein; intensity and duration of activity—activities more intense and long in duration will use more protein for fuel; degree of training—athletes use more protein as fuel than the unconditioned person. (9) Iron deficiency anemia is a condition that dramatically impairs physical performance. The athlete with this condition cannot use fat for fuel or perform aerobic activities, and therefore will tire easily. Sports anemia is a transient condition of low hemoglobin in the blood, associated with the early stage of sports training or other strenuous activity. It is an adaptive, temporary response to endurance training. Iron-deficiency anemia requires iron-supplementation, while sports anemia does not respond to such supplementation. (10) If the body loses too much water, its chemistry becomes compromised and dehydration may result, which can turn into heat stroke in hot humid weather. Recommendations are to drink enough water or diluted juice before and during training and competition, rest in the shade when tired, and wear lightweight clothing. (11) A nutrient-dense diet composed mostly of unprocessed foods that meets nutrient, fluid, and energy requirements.

Short Answers—

1. a. flexibility; b. strength; c. muscle endurance; d. cardiorespiratory endurance
2. a. aerobic; b. anaerobic
3. a. increased blood volume and oxygen delivery; b. increased heart strength and stroke volume; c. slowed resting pulse; d. increased breathing efficiency; e. improved circulation; f. reduced blood pressure

Crossword Puzzle—

```
              T R A I N I N G        H         A
              C         N            Y         T
              O         T            P         R
              N         E            E         O
              D U R A T I O N        R         P
              I         N   S E D E N T A R Y  H
              T         F   I        R
              I         I   T        O
              O         T   Y        P
              N         N            H
              I   F L E X I B I L I T Y
              N         S
              G   M U S C L E S T R E N G T H
```

Sample Test Questions—

1. e (p. 474)
2. e (p. 476)
3. b (p. 477)
4. c (p. 479)
5. b (p. 482)
6. d (p. 482)
7. b (p. 483)
8. a (p. 483)
9. b (p. 487)
10. c (p. 487)
11. e (p. 486-487)
12. c (p. 489)
13. a (p. 490)
14. b (p. 492)
15. b (p. 490)
16. c (p. 488)
17. a (p. 489)
18. c (p. 494)
19. e (p. 492)
20. d (p. 495)

Chapter 15
Life Cycle Nutrition: Pregnancy and Lactation

Chapter Outline

I. Nutrition Prior to Pregnancy

II. Growth and Development during Pregnancy
    A. Placental Development

    B. Fetal Growth and Development

    C. Critical Periods

III. Maternal Weight
    A. Weight Prior to Conception

    B. Weight Gain during Pregnancy

    C. Exercise during Pregnancy

IV. Nutrition during Pregnancy
    A. Energy and Nutrient Needs during Pregnancy

B. Common Nutrition-Related Concerns of Pregnancy

V. High-Risk Pregnancies
   A. The Infant's Birthweight

   B. Malnutrition and Pregnancy

   C. Food Assistance Programs

   D. Maternal Health

   E. The Mother's Age

   F. Practices Incompatible with Pregnancy

VI. Nutrition during Lactation
   A. Lactation: A Physiological Process

   B. Breastfeeding: A Learned Behavior

C. The Mother's Nutrient Needs

D. Practices Incompatible with Lactation

E. Maternal Health

Highlight: Fetal Alcohol Syndrome

SUMMING UP

(1)_____ is a major factor influencing the nutritional needs of developing infants and children. The growth rate is fastest during (2)_____ life and the first year. During pregnancy, changes in both mothers' and infants' bodies necessitate increased intakes of the (3)_____ nutrients. A pregnant woman should gain about (4)_____ pounds from foods of high nutrient density. Malnutrition during pregnancy affects the developing fetus; (5)_____-_____ babies often fail to thrive. (6)_____, smoking, drugs, dieting, and unbalanced nutrient intakes of all kinds should be avoided for the duration of pregnancy. (7)_____ and ample food energy are especially important. Fluid intake should be liberal and (8)_____ normally should not be restricted.

The breastfeeding mother needs additional (9)_____ from foods of high nutrient density and a generous fluid intake. The rapidly growing newborn infant requires milk, preferably (10)_____ milk, which provides the needed nutrients in quantities suitable to support the infant's growth. Advantages of breast milk over formula, especially in (11)_____ countries, are that it protects the infant against (12)_____ and that it is sanitary, economical, and premixed to the correct proportions. To avoid (13)_____, all susceptible infants should be breastfed at first, even in developed countries.

CHAPTER GLOSSARY

amniotic sac: the "bag of waters" in the uterus in which the fetus floats.
anencephaly: an uncommon and always fatal type of neural tube defect, characterized by the absence of a brain.
appropriate for gestational age (AGA): infants born at a size and weight appropriate for their gestational age.
cesarean section: a surgically assisted birth involving removal of the fetus by an incision into the uterus, usually by way of the abdominal wall.
colostrum: a milklike secretion from the breast, present during the first day or so after delivery before milk appears; rich in protective factors.
conception: the union of the male sperm and the female ovum; fertilization.
critical periods: finite periods during development in which certain events may occur that will have irreversible effects on later developmental stages. In a body organ, a critical period is usually a period of rapid cell division.
Down syndrome: a genetic abnormality that causes mental retardation, short stature and flattened facial features.
eclampsia: a severe stage of preeclampsia characterized by convulsions.
embryo: the developing infant from two to eight weeks after conception.
fertility: the capacity of a woman to produce a normal ovum periodically and of a man to produce normal sperm; the ability to reproduce.
fetal programming: the influences of substances during fetal growth on the development of diseases in later life.
fetus: the developing infant from eight weeks after conception until term.
food aversions: strong desires to avoid particular foods.
food cravings: strong desires to eat particular foods.
gestation: the period from conception to birth; for human beings gestation lasts from 38 to 42 weeks. Pregnancy is often divided into thirds, called *trimesters*.
gestational diabetes: the appearance of abnormal glucose tolerance that is first detected during pregnancy.
high-risk pregnancy: a pregnancy characterized by indicators that make it likely the birth will be surrounded by problems such as premature delivery, difficult birth, retarded growth, birth defects, and early infant death.
implantation: the stage of development in which the zygote embeds itself in the wall of the uterus and begins to develop; occurs during the first two weeks after conception.
lactation: production and secretion of breast milk for the purpose of nourishing an infant.
lactoferrin: a factor in breast milk that binds iron and keeps it from supporting the growth of the infant's intestinal bacteria.
let-down reflex: the reflex that forces milk to the front of the breast when the infant begins to nurse.
low birthweight (LBW): a birthweight of 5½ pounds (2500 grams) or less; indicates probable poor health in the newborn and poor nutrition status in the mother during pregnancy, before pregnancy, or both.
low-risk pregnancy: a pregnancy characterized by indicators that make a normal outcome likely.
macrosomia: high-birthweight infants resulting from prepregnancy obesity, excessive weight gain during pregnancy, or uncontrolled diabetes.

mammary glands: glands of the female breast that secrete milk.
milk anemia: iron-deficiency anemia that develops when an excessive milk intake displaces iron-rich foods from the diet.
neural tube defect: a serious central nervous system birth defect that often results in lifelong disability or death.
nursing bottle syndrome: extensive tooth decay due to prolonged tooth contact with formula, milk, fruit juice, or other carbohydrate-rich liquid offered to an infant in a bottle.
ovum: the female reproductive cell, capable of developing into a new organism upon fertilization; commonly referred to as an egg.
oxytocin: a hormone that stimulates the mammary glands to eject milk during lactation and the uterus to contract during childbirth.
placenta: the organ that develops inside the uterus early in pregnancy, in which maternal and fetal blood circulate in close proximity so that materials can be exchanged between them.
postpartum amenorrhea: the normal temporary absence of menstrual periods immediately following childbirth.
post term (infant): an infant born after the 42nd week of pregnancy.
preeclampsia: a condition characterized by hypertension, fluid retention, and protein in the urine.
pregnancy-induced hypertension (PIH): high blood pressure that develops in the second half of pregnancy.
preterm infant: an infant born prior to the 38th week of pregnancy; also called a *premature* infant. A term infant is born between the 38th and 42nd week of pregnancy.
prolactin: a hormone secreted from the anterior pituitary gland that acts on the mammary glands to initiate and sustain milk production.
small for gestational age (SGA): usually caused by malnutrition, this small infant has suffered growth failure in the uterus.
sperm: the male reproductive cell, capable of fertilizing an ovum.
spina bifida: one of the most common types of neural tube defects, characterized by the incomplete closure of the spinal cord and its bone encasement.
sudden infant death syndrome (SIDS): the unexpected and unexplained death of an apparently well infant; the most common cause of death of infants between the second week and the end of the first year of life; also called *crib death*.
teratogenic: causing abnormal fetal development and birth defects.
term infant: an infant born between the 38th and 42nd week of pregnancy.
toxemia: the hypertensive disease of pregnancy.
transient hypertension of pregnancy: high blood pressure that develops in the second half of pregnancy and resolves after childbirth.
umbilical cord: the ropelike structure through which the fetus's veins and arteries reach the placenta; the route of nourishment and oxygen into the fetus and the route of waste disposal from the fetus.
umbilicus: the scar in the middle of the abdomen that marks the former attachment of the umbilical cord, commonly known as the "belly button."
uterus: the muscular organ within which the infant develops before birth; the womb.
wean: to gradually replace breast milk with infant formula or other foods appropriate to an infant's diet.
zygote: the product of the union of ovum and sperm; so-called for the first two weeks after fertilization.

ASSIGNMENTS

Answer these chapter study questions from the text:

1. Describe the placenta and its function.

2. Describe the normal events of fetal development. How does malnutrition impair fetal development?

3. Define the term *critical period*. How do adverse influences during critical periods affect later health?

4. Explain why women of childbearing age need folate in their diets. How much is recommended, and how can women ensure that these needs are met?

5. What is the recommended pattern of weight gain during pregnancy for a woman at a healthy weight? For an underweight woman? For an overweight woman?

6. What does a pregnant woman need to know about exercise?

7. Which nutrients are needed in the greatest amounts during pregnancy? Why are they so important? Describe wise food choices for the pregnant woman.

8. Define low-risk and high-risk pregnancies. What is the significance of infant birthweight in terms of the child's future health?

9. Describe some of the special problems of the pregnant adolescent. Which nutrients are needed in increased amounts?

10. What practices should be avoided during pregnancy? Why?

11. How do nutrient needs during lactation differ from nutrient needs during pregnancy?

Complete this short answer question:

1. List conditions that raise risk in pregnancy:

   a.

   b.

   c.

   d.

   e.

   f.

   g.

Complete this crossword puzzle by Mary A. Wyandt, Ph.D., CHES.

Across:	Down:
1. the period from conception to birth; for humans it lasts from 38 to 42 weeks	2. the female reproductive cell
4. the male reproductive cell	3. the developing infant from eight weeks after conception until term
7. the muscular organ within which the infant develops before birth; the womb	5. the organ that develops inside the uterus early in pregnancy, in which maternal and fetal blood circulate in close proximity so that materials can be exchanged between them
9. the capacity of a woman to produce a normal ovum periodically and of a man to produce normal sperm; the ability to reproduce	6. the product of the union of ovum and sperm; so-called for the first two weeks after fertilization
10. the union of the male sperm and the female ovum; fertilization	8. the developing infant from two to eight weeks after conception

Solve these problems:

1. What is the recommended energy intake during pregnancy for a woman who weighs 130 lb.?

2. Ideally, how much weight would you expect a woman to have gained by the 32nd week of pregnancy?

SAMPLE TEST QUESTIONS

1. Nutrients and oxygen travel to the developing fetus via:

    a. the placenta.
    b. the amniotic sac.
    c. its lungs.
    d. its intestines.

2. The growth period when a lack of nutrients is most likely to produce permanent change is the:

    a. period of rapid increase in number of cells.
    b. period of rapid increase in size of cells.
    c. teen growth spurt.
    d. growth spurt from 1 to 20 years of age.

3. A fetus is the:

    a. developing infant during its second through eighth week after conception.
    b. developing infant from the eighth week after conception until birth.
    c. muscular organ within which the infant develops before birth.
    d. infant from birth until its first birthday.
    e. organ from which the infant receives nourishment.

4. A pregnant woman (compared to a non-pregnant woman) is advised to consume almost twice as much/many:

    a. kcalories.
    b. iron.
    c. fat.
    d. carbohydrate.
    e. folate.

5. During pregnancy, physicians often recommend:

    a. weight gain of about 25 to 35 pounds.
    b. no mineral or vitamin supplements and weight gain of at least 35 pounds.
    c. vitamin C supplements and weight gain of about 20 pounds.
    d. multi-vitamin and mineral supplements and weight gain of no more than 20 pounds.

6. Because of the dangers of obesity, an overweight woman should try to gain little or no weight during pregnancy.

   a.   True   b.   False

7. The rate of weight gain during pregnancy for normal-weight women should optimally be:

   a.   1 pound per week in the 1st trimester and 1 pound per month thereafter.
   b.   10 pounds in the 1st trimester and 1 pound per week thereafter.
   c.   3 ½ pounds in the 1st trimester and just under 1 pound per week thereafter.
   d.   no weight until at least the 2nd trimester, then 1 pound per week.

8. Damage done to an infant by a woman's alcohol abuse during pregnancy can be corrected after birth by proper nutrition.

   a.   True   b.   False

9. A low-birthweight baby has a statistically greater chance of contracting diseases and of dying early in life.

   a.   True   b.   False

10. The stage of development in which the fertilized egg embeds itself in the wall of the uterus is called:

    a.   ovum.   d.   zygote.
    b.   critical period.   e.   implantation.
    c.   placenta.

11. A pregnant woman who is constipated asks a health care provider for advice. Assuming no disease is present, what should the health care provider probably tell her to try first?

    a.   increase fluid intake   c.   eat fewer bulky foods
    b.   use laxatives   d.   reduce fluid intake

12. A pregnant woman needs _____ kcalories above the allowance for a nonpregnant woman.

    a.   300   c.   500
    b.   400   d.   600

13. The protein RDA for pregnancy is _____ grams per day higher than for nonpregnant women.

    a.   2
    b.   6
    c.   10
    d.   25

14. The increased need for this nutrient in pregnancy is difficult to meet by diet or by existing stores; therefore supplements are recommended.

    a. vitamin C
    b. sodium
    c. iron
    d. vitamin D

15. The most common outcome of a high-risk pregnancy is:

    a. diabetes.
    b. heart disease.
    c. cancer.
    d. low birthweight.

16. Pregnant teenagers have higher rates of

    a. maternal deaths.
    b. preterm births.
    c. low birthweight infants.
    d. all of the above
    e. b and c

ANSWERS

Summing Up—(1) Growth; (2) prenatal; (3) growth; (4) 25 to 30; (5) low-birthweight; (6) Alcohol; (7) Protein; (8) salt; (9) kcalories; (10) breast; (11) developing; (12) disease; (13) allergy.

Chapter study questions from the text—(1) The placenta is the organ that develops inside the uterus early in pregnancy, in which maternal and fetal blood circulate in close proximity so that materials can be exchanged between them. The baby receives nutrients and oxygen across the placenta, and the mother's blood picks up carbon dioxide and other waste products to be excreted. (2) During the 7 months of fetal development, each organ grows to maturity according to its own characteristic schedule, with greater intensity at some times than at others. Intense development and rapid cell division take place only during certain times. During times of critical periods each organ and tissue is most vulnerable to an insult such as a nutrient deficiency. If nutrients are not available during these times, significant damage can take place with regard to organ development. (3) Critical periods are finite periods during development in which certain events may occur that will have irreversible effects on later developmental stages. In the case of the developing fetus, a critical period is a period of rapid cell division. Early malnutrition during these times affects the heart and lungs in later life. (4) Folate can prevent neural tube defects. 0.4 mg/day; eat plenty of fruits and vegetables or take supplements. (5) Underweight: 28 to 40 lb.; normal-weight: 25 to 35 lb.; overweight: 15 to 24 lb. (6) A woman who is active prior to pregnancy and is experiencing a normal pregnancy can continue to exercise throughout her pregnancy, but will want to avoid sports in which she might fall or be hit by other people or objects such as racquetball. Women are advised to avoid overly strenuous activities and to abstain from exercising in hot weather, and to stay out of saunas, steam rooms, and hot whirlpools. (7) Energy nutrients are essential (a women needs extra food energy to support the growth and metabolic activities of the placenta, fetus, and her own body tissues), extra protein is needed for growth of new tissues, additional B vitamins are needed in proportion to increased food energy intake (they function as coenzymes in energy reactions), folate and B_{12} are needed for blood cells and growth, vitamin D and minerals are necessary for bone development, extra iron is needed to provide for fetal and placental needs and usually to boost pre-pregnancy low blood levels, and zinc is needed for DNA and RNA synthesis (thus for protein synthesis and cell development). Wise food choices include: a balance similar to that of the Daily Food Guide (Ch. 2) with additional servings for each of the five food groups. (8) Low-risk pregnancy: a pregnancy characterized by indicators that make a normal

outcome likely; high-risk pregnancy: a pregnancy characterized by indicators that make it likely the birth will be surrounded by problems such as premature delivery, difficult birth, retarded growth, birth defects, and early infant death. Birthweight is a potential predictor of an infant's future health and survival. (9) A teenaged female is growing and has to meet her own nutrient needs, in addition to trying to meet the nutrient needs of a developing baby. The demands of pregnancy compete with those of her own growth, placing her and the infant at high risk for complications. Protein and vitamin K.
(10) Drinking alcohol, taking medicinal drugs, using illicit drugs, smoking or using smokeless tobacco, ingesting foods and beverages contaminated with lead or mercury, taking vitamin-mineral megadoses, drinking caffeine, weight-loss dieting. Substances can pass to the fetus and cause irreversible harm or death. (11) In general, nutrient needs are as high or higher during lactation as during pregnancy (except protein and folate).

Short Answer—
1. alcohol consumption; use of medicinal drugs, herbal supplements, illicit drugs, tobacco; exposure to environmental contaminants; experiencing foodborne illness; consuming vitamin/mineral megadoses; consuming caffeine; weight loss dieting

Crossword Puzzle—

Across/Down answers filled in grid: GESTATION, SPERM, UTERUS, FERTILITY, CONCEPTION, PLACENTA, ZYGOTE, FETUS, EMBRYO, OVUM

Problem Solving—
1. 130 lb. X 18 kcal/lb. = 2,340 kcal or 59 kg X 40 kcal/kg = 2,360 kcal
2. 2 to 4 lb./1st 3 mo (12 wk); 32 wk - 12 wk = 20 wk; 2 to 4 lb. + 20 lb. (1 lb./wk) = 22 to 24 lb.

Sample Test Questions—
1. a (p. 508)
2. a (p. 510)
3. b (p. 510)
4. e (p. 512)
5. a (p. 514)
6. b (p. 514)
7. c (p. 515)
8. b (p. 527-528)
9. a (p. 514)
10. e (p. 508)
11. a (p. 522)
12. a (p. 518)
13. d (p. 518)
14. c (p. 520)
15. d (p. 523)
16. d (p. 527)

Chapter 16
Life Cycle Nutrition: Infancy, Childhood, and Adolescence

Chapter Outline

I. Nutrition during Infancy
   A. Energy and Nutrient Needs

   B. Breast Milk

   C. Infant Formula

   D. Special Needs of Preterm Infants

   E. Introducing Cow's Milk

   F. Introducing Solid Foods

   G. Mealtimes with Toddlers

II. Nutrition during Childhood
   A. Energy and Nutrient Needs

B. Hunger and Malnutrition in Children

C. The Malnutrition-Lead Connection

D. Hyperactivity and "Hyper" Behavior

E. Food Allergy and Intolerance

F. Childhood Obesity

G. Mealtimes at Home

H. Nutrition at School

III. Nutrition during Adolescence
    A. Growth and Development

    B. Energy and Nutrient Needs

C. Food Choices and Health Habits

_____
_____
_____

D. Problems Adolescents Face

_____
_____
_____

Highlight: Childhood Obesity and the Early Development of Chronic Diseases

_____
_____
_____

SUMMING UP

After the age of one, a child's growth rate (1)_____, and with it, the appetite. However, all essential nutrients continue to be needed in adequate amounts from foods with a high nutrient (2)_____.

When children go to school, their nutrition needs are partly met by school (3)_____ programs. Another influential factor in the lives of children is (4)_____, with many advertisements for sugary foods; another is (5)_____ machines, which often limit choices to foods of low quality. (6)_____ and other health professionals are concerned that the advertisement and availability of sugary foods should be controlled; there may also be a need to control some children's (7)_____ consumption from cola beverages and cocoa products.

Sound nutrition practices may prevent future health problems to some extent—among them, (8)_____, iron-deficiency anemia, (9)_____ disease, and diabetes. Screening for these conditions ensures early detection and facilitates early control.

It is desirable for children to learn to like (10)_____ foods in all the food groups. This liking seems to come naturally except, in some children, the liking for (11)_____. The person who feeds the child must be aware of the child's (12)_____ and emotional development. Children can learn positive eating (13)_____ that will continue to promote their good health after they have become adults.

The (14)_____ years mark the transition from a time when children eat what they are fed to a time when they choose for themselves what to eat. Nutrition (15)_____ becomes important as a means of encouraging healthy food habits. Teenagers' (16)_____ patterns and lifestyles predispose them to certain nutrient inadequacies, notably a lack of (17)_____, but teenagers vary so widely that generalizations are difficult. Special problems that may arise in the teen years and affect nutrition include pregnancy, alcohol, tobacco and (18)_____ use.

CHAPTER GLOSSARY

adolescence: the period from the beginning of puberty until maturity.
adverse reactions: unusual responses to food (including intolerances and allergies).
alpha-lactalbumin: the chief protein in human breast milk, as opposed to *casein*, the chief protein in cow's milk.
anaphylactic shock: a life-threatening whole-body allergic reaction.
asymptomatic allergy: producing antibodies without symptoms.
beikost: supplemental, or weaning foods.
bifidus factors: factors in colostrum and breast milk that favor the growth of the "friendly" bacteria *Lactobacillus bifidus* in the infant's intestinal tract, so that other, less desirable intestinal inhabitants will not flourish.
botulism: an often fatal food-borne illness caused by the ingestion of foods containing a toxin produced by bacteria that grow without oxygen.
colostrum: a milklike secretion from the breast, present during the first day or so after delivery before milk appears; rich in protective factors.
dental caries: decay of the teeth.
epinephrine: a hormone of the adrenal gland administered by injection to counteract anaphylactic shock by opening the airways and maintaining heartbeat and blood pressure.
food allergy: an adverse reaction to foods that involves an immune response; also called *food-hypersensitivity reaction*.
food intolerances: adverse reactions to foods that do not involve the immune system.
gatekeepers: with respect to nutrition, key people who control other people's access to foods and thereby exert profound impacts on their nutrition.
hyperactivity: inattentive and impulsive behavior that is more frequent and severe than is typical of others at a similar age; professionally called *attention deficit hyperactivity disorder (ADHD)*.
hypoallergenic formulas: clinically tested infant formulas that do not provoke reactions in 90% of infants or children with confirmed cow's milk allergy.
lactadherin: a protein in breast milk that attacks diarrhea-causing viruses.
lactoferrin: a factor in breast milk that binds iron and keeps it from supporting the growth of the infant's intestinal bacteria.
milk anemia: iron-deficiency anemia that develops when an excessive milk intake displaces iron-rich foods from the diet.
nursing bottle tooth decay: extensive tooth decay due to prolonged contact with formula, milk, fruit juice, or other carbohydrate-rich liquid offered to an infant in a bottle.
osteopenia: a metabolic bone disease common in preterm infants; also called *rickets of prematurity*.
puberty: the period in life in which a person becomes physically capable of reproduction.
symptomatic allergy: producing antibodies and symptoms.
wean: to gradually replace breast milk with infant formula or other foods appropriate to an infant's diet.

ASSIGNMENTS

Complete this crossword puzzle by Mary A. Wyandt, Ph.D., CHES.

Across:	Down:
1. a protein in breast milk that attacks diarrhea-causing viruses	2. the chief protein in human breast milk
7. adverse reactions to foods that do not involve the immune system	3. a factor in breast milk that binds iron and keeps it from supporting the growth of the infant's intestinal bacteria
8. a milk-like secretion from the breast, present during the first day or so after delivery before milk appears; rich in protective factors	4. supplemental or weaning foods
9. an often fatal food-borne illness caused by ingesting foods that contain a toxin produced by bacteria that grow without oxygen	5. to gradually replace breast mild with infant formula or other foods appropriate to an infant's diet
10. a metabolic bone disease common in preterm infants	6. the period from the beginning of puberty until maturity

Answer these chapter study questions from the text:

1. Describe some of the nutrient and immunological attributes of breast milk.

2. What are the appropriate uses of formula feeding? What criteria would you use in selecting an infant formula?

3. Why are solid foods not recommended for an infant during the first few months of life? When is an infant ready to start eating solid food?

4. Identify foods that are inappropriate for infants and explain why they are inappropriate.

5. What nutrition problems are most common in children? What strategies can help prevent them?

6. Describe the relationships between nutrition and behavior. How does television influence nutrition?

7. Describe a true food allergy. Which foods most often cause allergic reactions? How do food allergies influence nutrition status?

8. Describe the problems associated with childhood obesity and the strategies for prevention and treatment.

9. List strategies for introducing nutritious foods to children.

10. What impact do school meal programs have on the nutrition status of children?

11. Describe changes in nutrient needs from childhood to adolescence. Why is a teenaged girl more likely to develop an iron deficiency than is a boy?

12. How do adolescents' eating habits influence their nutrient intakes?

13. How does the use of illicit drugs influence nutrition status?

14. How do the nutrient intakes of smokers differ from those of nonsmokers? What impacts can those differences exert on health?

Complete these short answer questions:

1. List iron-rich foods that most children like.

2. List 5 physical signs of malnutrition in children.

    a.

    b.

    c.

    d.

    e.

SAMPLE TEST QUESTIONS

1. After the age of one, a child's growth rate accelerates.

    a. True   b. False

2. The American Academy of Pediatrics recommends that infants receive breast milk for the first 6 to 12 months.

    a. True   b. False

3. At what age does the normal infant first develop the ability to swallow solid food?

    a. 3-5 weeks   c. 4-6 months
    b. 26-32 weeks   d. 9-12 months

4. What should be the first food introduced to the infant?

    a. yogurt
    b. egg white
    c. rice cereal
    d. finely chopped meat

5. The main purpose of introducing solid food is:

   a. to help the infant sleep through the night.
   b. to provide nutrients that are no longer supplied adequately by breast milk alone.
   c. to increase body weight.
   d. to improve mental capacity.

6. Which of the following changes in body structure usually takes place between the ages of 1 and 2 years?

   a. weight doubles
   b. weight triples
   c. length of body doubles
   d. length of long bones increases

7. Approximately how many kcals per day does an average 3-year old need to obtain?

   a. 500
   b. 800
   c. 1300
   d. 2400

8. Which of the following provides the most important information about a child's health?

   a. growth chart
   b. blood lipid profile
   c. long-bone size and density
   d. onset of walking and talking

9. Two to six year olds need _____ servings from the milk group daily.

   a. one
   b. one-half
   c. two
   d. three

10. A large amount of concentrated sweets in a child's diet is most likely to lead to:

    a. apathy.
    b. obesity.
    c. hyperactivity.
    d. growth inhibition.

11. An adverse reaction to foods that does not involve an immune response is:

    a. food intolerance.
    b. food allergy.
    c. antigen.
    d. anaphylactic.

12. Hyperactivity is caused by a high sugar diet.

    a. True
    b. False

13. How much more total energy does a normal 10 year old need compared with a normal 1 year old?

    a. 25%
    b. 50%
    c. 100%
    d. 200%

14. Which of the following two conditions are associated with television's influence?

    a. obesity and poor dental health
    b. drug abuse and teenage pregnancy
    c. anorexia and nutrient deficiencies
    d. hyperactivity and lower body weight

15. Compared with adolescent boys in terms of height, weight, and body composition, adolescent girls:

    a. start growing later.
    b. lay down more fat.
    c. have more lean body mass.
    d. have more energy.

16. The School Lunch Program is intended to provide at least _____ of children's RDA for each nutrient.

    a. one fourth
    b. one third
    c. one hundred percent
    d. No requirement stipulated.

17. One nutrient that often comes up short in teenagers' diets is:

    a. protein.
    b. potassium.
    c. thiamin.
    d. iron.
    e. riboflavin.

18. Nutrients often found lacking in teenage snacks include all but one of the following:

    a. iron.
    b. calcium.
    c. fat.
    d. vitamin A.
    e. folate.

19. The single most effective way to teach nutrition to children is by:

    a. example.
    b. punishment.
    c. singling out only hazardous nutrition practices for attention.
    d. explaining the importance of eating new foods as a prerequisite for dessert.

20. What can parents do to help their children consume a balanced diet?

    a. Allow eating only at mealtimes and forbid all snacking.
    b. Make a variety of nutritious foods available.
    c. Insist on keeping close track of everything they eat.
    d. Parents really can't do anything to influence their diets.

21. A teenager's kcalorie intake from snacks averages about _____ percent of total daily intake.

    a. five
    b. fifteen
    c. twenty-five
    d. forty
    e. fifty

22. A nutrition problem observed in people with drug addictions is:

    a. they spend money for drugs that could be spent on food.
    b. they lose interest in food during highs.
    c. some drugs depress the appetite.
    d. all of the above.

ANSWERS

Summing Up—(1) slows; (2) density; (3) lunch; (4) television; (5) vending; (6) Dentists; (7) caffeine; (8) obesity; (9) cardiovascular; (10) nutritious; (11) vegetables; (12) psychological; (13) habits; (14) teen; (15) education; (16) snacking; (17) iron; (18) drug.

Chapter study questions from the text—(1) Breast milk provides natural antibodies, it is sterile, and it is nutritionally tailor-made to meet infant's needs. (2) To substitute for breast milk occasionally, or to wean to formula during the first year. Must meet AAP standards. (3) Allergy development, physical immaturity, nutrient needs met by stores and formula or breast milk until then. Infant is ready to start eating solid food when the infant's birth weight has doubled, the infant consumes 8 ounces of formula and gets hungry again in less than 4 hours, the infant can sit up, the infant consumed 32 ounces a day and wants more, the infant is six months old. (4) Ordinary milk because it provides insufficient vitamin C and iron and excessive sodium and protein; mixed dinners and heavily sweetened desserts because they are not nutrient dense; sweets of any kind including baby food "desserts" because they convey no nutrients to support growth and may promote obesity; canned vegetables because they contain too much sodium, honey and corn syrup because of the risk of botulism; popcorn, whole grapes, whole beans, hot dog slices, hard candies and nuts because they can easily choke on these foods. (5) Iron-deficiency anemia; children should receive 5.5 mg of iron per 1,000 kcalories. Enough milk must be provided to ensure adequate calcium and riboflavin intake but no more. After age 2, low fat milk should be used so there will be more calories available for iron rich foods such as lean meats, fish, poultry, eggs and legumes.
(6) Nutrient deficiencies incur behavioral symptoms. Iron deficiency causes an energy crisis and directly affects moods, attention span, and learning ability. Television commercials promote sugary foods. (7) A true food allergy is an adverse reaction to foods that involves an immune response (common allergic reactions are skin rash, digestive upset, or respiratory discomfort); eggs, peanuts, and milk are likely to cause allergy. Food allergies can influence a person's nutrition status in that when trying to identify a suspected food, the food is omitted from the diet and this invites the risk of nutrient deficiencies.
(8) Obese children are most likely to become obese adults and therefore are at risk for the social, economic and medical ramifications that often accompany obesity. Obese children begin puberty earlier and grow taller than their peers at first, but stop growing at a short height; they display higher levels of total cholesterol, triglycerides, LDL and VLDL; they tend to have high blood pressure; they are at risk for diabetes, asthma; they are victims of prejudice. Strategies include modifying dietary and exercise patterns. The dietary goals are to reduce the rate of weight gain; that is, to maintain weight while growing in height. Children are encouraged to eat slowly and select nutrient dense foods and not be pressured to clean their plates. Daily physical activity should be included. (9) Parents can allow children to select from

healthful choices, to prepare foods, to grow foods in a garden, to visit food-related places. (10) School lunches can contribute to positive nutrition status if they are nutritious and acceptable. (11) Nutrient needs increase during adolescence because it is a time of growth. Energy needs increase dramatically especially for males, iron needs increase especially for females because of monthly blood loss, and calcium needs increase for both genders. (12) Adolescents almost inevitably fall into irregular eating habits and snacks provide about a fourth of the average teenager's total daily food energy intake. Many snack foods offer a variety of nutrients although they are usually high in kcalories and low in vitamin A, folate, fiber and sometimes calcium. (13) Marijuana sometimes alters the sense of taste and increases enjoyableness of eating; prolonged use of the drug does not seem to increase energy intake enough to bring about weight gain. Cocaine causes lack of appetite and often addiction. Users often spend money on the drug instead of food, and lose interest in food during times they are using the drug. Alcohol is an empty-kcalorie beverage that can displace needed nutrients from the diet while simultaneously altering absorption and metabolism of nutrients. Tobacco smokers have lower intakes of dietary fiber, vitamin A, folate and vitamin C. (14) Smokers tend to have lower intakes of dietary fiber, vitamin A, beta-carotene, folate, and vitamin C. Since smokers require higher levels of some nutrients, their low intakes may increase their risks of cancer.

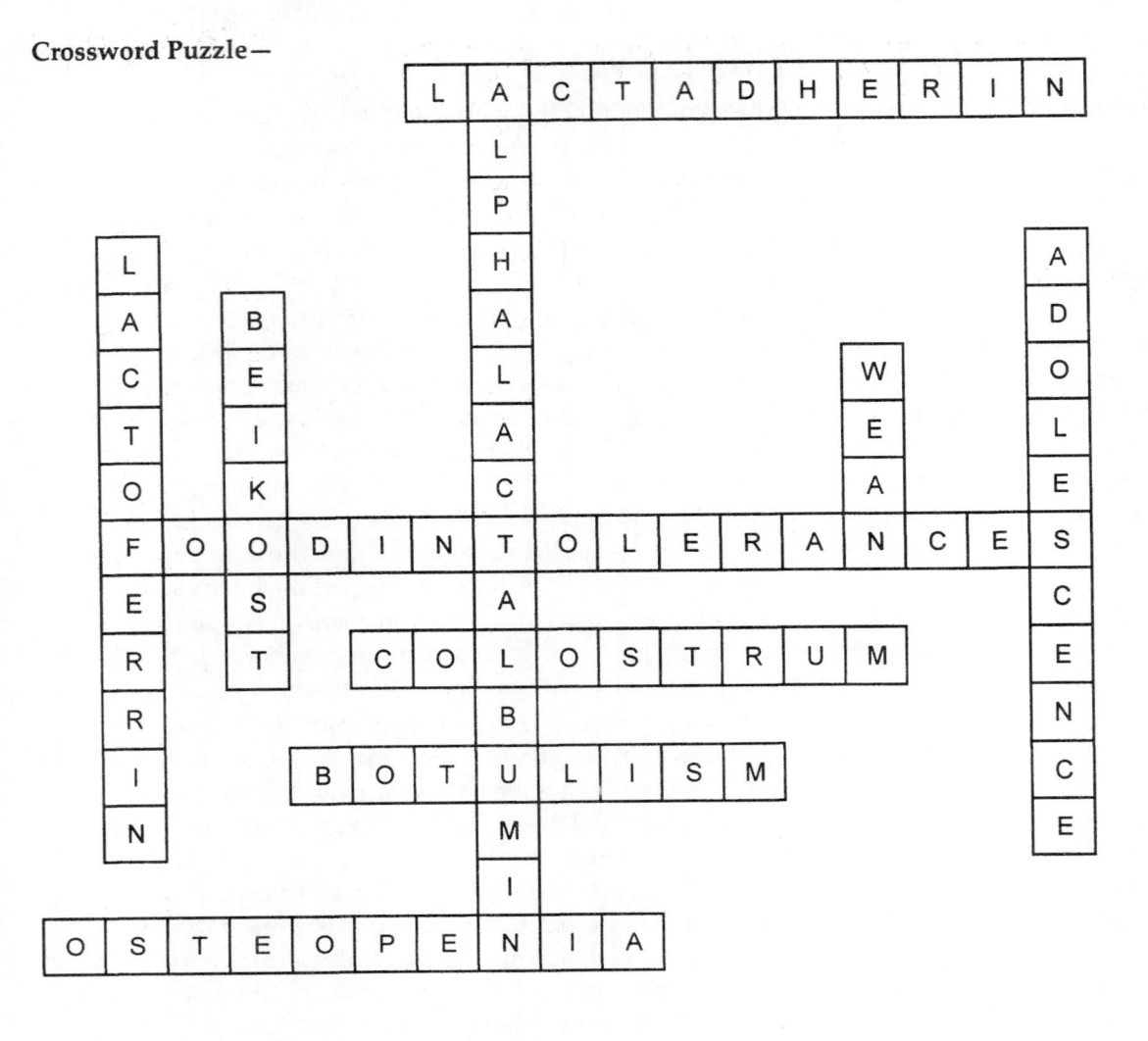

Crossword Puzzle—

Across/Down answers filled in:
- LACTADHERIN
- FOOD INTOLERANCES
- COLOSTRUM
- BOTULISM
- OSTEOPENIA
- LACTOFERRIN
- BEIKOST
- ALPHALACTALBUMIN
- WEANING
- ADOLESCENCE

Short Answers—

1. Canned macaroni, canned spaghetti, cream of wheat, fortified dry cereals, noodles, rice or barley, tortillas, whole-wheat, enriched, or fortified bread, bran muffins, baked potato skins, cooked mushrooms, cooked mung bean sprouts or snow peas, green peas, mixed vegetable juice, canned plums, cooked dried apricots, dried peaches, raisins, bean dip, canned pork and beans, mild chilli or other bean-meat dishes such as burritos, liverwurst, meat casseroles, peanut butter and jelly sandwich, lean roast beef or cooked ground beef, sloppy joes.
2. See Table 16-3.

Sample Test Questions—

1.	b (p. 543)	9.	c (p. 557)	17.	d (p. 572)
2.	a (p. 546)	10.	b (p. 553)	18.	c (p. 572)
3.	c (p. 551)	11.	a (p. 562)	19.	a (p. 567-568)
4.	c (p. 551)	12.	b (p. 560)	20.	b (p. 570)
5.	b (p. 551)	13.	c (p. 555)	21.	c (p. 572)
6.	d (p. 554)	14.	a (p. 564)	22.	d (p. 574)
7.	c (p. 555)	15.	b (p. 571)		
8.	a (p. 554)	16.	b (p. 569)		

CHAPTER 17
LIFE CYCLE NUTRITION: ADULTHOOD AND THE LATER YEARS

CHAPTER OUTLINE

I. Nutrition and Longevity
    A. Observation of Elderly People

    B. Manipulation of Diet

II. The Aging Process
    A. Physiological Changes

    B. Other Changes

III. Energy and Nutrient Needs of Older Adults
    A. Water

    B. Energy and Energy Nutrients

    C. Vitamins and Minerals

    D. Nutrient Supplements

IV. Nutrition-Related Concerns of Older Adults
   A. Cataracts and Macular Degeneration

   _____
   _____
   _____
   _____

   B. Arthritis

   _____
   _____
   _____
   _____

   C. The Aging Brain

   _____
   _____
   _____
   _____

V. Food Choices and Eating Habits of Older Adults
   A. Nutrition Assistance Programs

   _____
   _____
   _____
   _____

   B. Meals for Singles

   _____
   _____
   _____
   _____

Highlight: Nutrient-Drug Interactions

_____
_____
_____
_____

SUMMING UP

Adequate nutrition can help an individual reach the maximum life span by postponing and slowing (1)_____. Nutrition affects aging by its role in disease prevention. Six healthy habits to profoundly affect aging: abstinence from, or moderation in, (2)_____ use; regularity of meals; (3)_____ control; regular, adequate (4)_____; abstinence from (5)_____; and regular (6)_____ _____. Some physiological changes that occur with age include: loss of (7)_____ and lean body mass; decline in (8)_____ system function; slow GI motility; and (9)_____ loss.

Setting nutrient standards for older people is difficult because (10)_____ differences become more pronounced as people age, and (11)_____ needs decline with advancing age. Other nutrient needs do not vary significantly with those of adults. However, elderly people often have problems obtaining the (12)_____ they need. Nutrition may provide some protection against some of the conditions associated with aging including cataracts and (13)_____. Food assistance programs provide (14)_____ meals to older adults.

Chapter Glossary

Alzheimer's disease: a degenerative disease of the brain involving memory loss and major structural changes in neuron networks; also known as primary degenerative dementia of senile onset or chronic brain syndrome, also known as *senile dementia of the Alzheimer's type (SDAT)*.
arthritis: inflammation of a joint usually accompanied by pain, swelling, and structural changes.
atrophic gastritis: chronic inflammation of the stomach accompanied by a diminished size and functioning of the mucosa and glands.
cataracts: thickenings of the eye lenses that impair vision and can lead to blindness.
chronological age: a person's age in years from his or her date of birth.
congregate meals: nutrition programs that provide food for the elderly in a conveniently located setting such as a community center.
dysphagia: difficulty in swallowing.
edentulous: lack of teeth.
life expectancy: the average number of years lived by people in a given society.
life span: the maximum number of years of life attainable by a member of a species.
longevity: long duration of life.
macular degeneration: deterioration of the macular area of the eye that can lead to loss of central vision and eventual blindness.
Meals on Wheels: a nutrition program that delivers food for the elderly to their homes.
neurons: nerve cells; the structural and functional units of the nervous system. Neurons initiate and conduct nerve transmissions.
neurotransmitter: a chemical agent released by one neuron that acts upon a second neuron or upon a muscle or gland cell and alters its electrical state or activity.
osteoarthritis: a painful, chronic disease of the joints that occurs when the cushioning cartilage in a joint breaks down; joint structure is usually altered, with loss of function; also called **degenerative arthritis.**
quality of life: a person's perceived physical and mental well being.
physiological age: a person's age as estimated from her or his body's health and probable life expectancy.
pressure ulcers: damage to the skin and underlying tissues as a result of compression and poor circulation; commonly seen in people who are bedridden.
rheumatoid arthritis: a disease of the immune system involving painful inflammation of the joints and related structures.
sarcopenia: loss of skeletal muscle mass, strength, and quality.
senile dementia: the loss of brain function beyond the normal loss of physical adeptness and memory that occurs with aging.
stress: any threat to a person's well-being; a demand placed on the body to adapt.
stressors: environmental elements, physical or psychological, that cause stress.

stress response: the body's response to stress, mediated by both nerves and hormones.
ultrahigh temperature (UHT): a process used to treat milk so that it can be stored at room temperature.

ASSIGNMENTS

Answer these chapter study questions from the text:

1. What roles does nutrition play in aging, and what roles can it play in retarding aging?

2. What are some of the physiological changes that occur in the body's systems with aging? To what extent can aging be prevented?

3. Why does the risk of dehydration increase as people age?

4. Why do energy needs usually decline with advancing age?

5. Which vitamins and minerals need special consideration for the elderly? Explain why. Name some factors that complicate the task of setting nutrient standards for older adults.

6. Discuss the relationships between nutrition and cataracts and between nutrition and arthritis.

7. What characteristics contribute to malnutrition in older people?

Complete this crossword puzzle by Mary A. Wyandt, Ph.D., CHES.

Across:	Down:
3. any threat to a person's well-being; a demand placed on the body to adapt	1. nutrition programs that provide food for the elderly in a conveniently located setting such as a community center
7. chronic inflammation of the stomach accompanied by a diminished size and functioning of the mucosa and glands	2. a chemical agent released by one neuron that acts upon a second neuron or upon a muscle or gland cell and alters its electrical state or activity
8. environmental elements, physical or psychological, that cause stress	4. lack of teeth
9. long duration of life	5. nerve cells; the structural and functional units of the nervous system
10. loss of skeletal muscle mass, strength, and quality	6. difficulty in swallowing

Complete this short answer question:

1. List 7 strategies for growing old healthfully.

   a.

   b.

   c.

   d.

   e.

   f.

   g.

SAMPLE TEST QUESTIONS

1. What is the life expectancy for white males and females in the U. S.?

   a. 65, 70 years   c. 79, 84 years
   b. 75, 80 years   d. 85, 89 years

2. Studies of adults show that longevity is related, in part, to all of the following *except*:

   a. weight control.
   b. regularity of meals.
   c. short periods of sleep.
   d. no or moderate alcohol intake.

3. What would be the physiological age of a 75 year old woman whose physical health is equivalent to that of her 50 year old daughter?

   a. 25 years   c. 70 years
   b. 50 years   d. 125 years

4. Which of the following has been associated with regular physical activity in older adults?

   a. retention of sodium
   b. increase in blood LDL
   c. increased death rates
   d. slowing of cardiovascular aging

5. All of the following environmental factors are known to promote aging *except*:

   a. exercise.
   b. diseases.
   c. lack of nutrients.
   d. extremes of heat and cold.

6. As people age, they often experience:

   a. declining immune system function.
   b. loss of elasticity of intestinal wall.
   c. tooth loss.
   d. all of the above
   e. a and b

7. Atrophic gastritis is a condition that can especially impair the absorption of the following nutrients:

   a. vitamin B₁₂ and biotin.
   b. calcium and iron.
   c. glucose and amino acids.
   d. all of the above
   e. a and b

8. Studies of the eating habits of older adults demonstrate all of the following *except*:

   a. those who live alone in federally funded housing had higher quality diets.
   b. adults living alone consume insufficient amounts of food.
   c. malnutrition was associated with a lower level of education.
   d. men living with spouses ate higher quality diets than men living alone.

9. Nutrient needs of older people:

   a. vary according to individuals and show pronounced differences.
   b. increase; therefore, supplementation is required.
   c. remain the same as in young adult life.
   d. decrease.

10. Which of the following is a feature of elderly people and water metabolism?

    a. they do not feel thirsty or recognize dryness of the mouth
    b. they have a higher total body water content compared with younger adults
    c. they show increased frequency of urination which results in higher requirements
    d. they frequently show symptoms of overhydration such as mental lapses and disorientation

11. Which of the following statements describes one aspect of mineral nutrition of older adults?

    a. zinc intake is adequate for about 95% of this group
    b. calcium intakes of females are near the RDA for this group
    c. iron-deficiency anemia in this population group is less common than in younger adults
    d. calcium allowances for this group have recently been increased by the Committee on Dietary Allowances

12. What are the thickenings that occur to the lenses of the eye, thereby affecting vision, especially in the elderly?

    a. keratoids
    b. cataracts
    c. retinitis
    d. rhodopolids

13. What nutrients may be protective against cataract formation?

    a. iron and calcium
    b. chromium and zinc
    c. vitamin B$_{12}$ and folate
    d. vitamin C and vitamin E

14. How is nutrition linked to rheumatoid arthritis?

    a. the immune system relies on adequate nutrition
    b. omega-3 fatty acids may interfere with the action of prostaglandins
    c. a and b
    d. none of the above

15. Goals of the federal Nutrition Program for the elderly include the provision of all of the following *except*:

    a. transportation services.
    b. high-cost nutritious meals.
    c. opportunity for social interaction.
    d. counseling and referral to other social services.

Answers

Summing Up—(1) diseases; (2) alcohol; (3) weight; (4) sleep; (5) smoking; (6) physical activity; (7) bone; (8) immune; (9) tooth; (10) individual; (11) Energy; (12) nutrients; (13) arthritis; (14) nutritious.

Chapter study questions from the text—(1) Nutrition can slow some aspects of the aging process within the natural limits set by heredity. Eating well can help prevent diseases common in old age (diabetes, obesity and CVD). Eating well can prevent deficiency diseases. (2) Energy needs decrease; therefore, all foods must be nutrient dense. Protein needs are about the same or slightly greater than during young adult age. Aging cannot be prevented. (3) Risk of dehydration increases as people age because older people do not feel thirsty or notice mouth dryness. (4) Energy needs usually decline with advancing age because lean body mass diminishes, reducing basal metabolic rate. As people age they reduce their physical activity. (5) Most vitamin and mineral needs remain constant from early adulthood to later years; however, vitamin A needs special consideration because processing vitamin A within the body slows slightly. Vitamin D deficiency is possible because most adults drink little or no milk. Iron deficiency anemia occurs in many older adults. Zinc intake remains low in this population. Researchers face challenges in setting nutrient standards for older adults because the accuracy of dietary intake studies is difficult, because the amounts of certain minerals in foods is unknown. Interactions with other nutrients and drugs alter the bioavailability of some minerals; age-related metabolic changes and disease conditions make the task even more difficult. (6) Oxidative stress plays a role in cataract development and antioxidant nutrients may help minimize the damage; consuming adequate intakes of vitamins C and E and carotenoids is important. Being overweight aggravates arthritis partly because weight-bearing joints have to carry excess poundage. (7) Physically older people may not be mobile and as able to purchase and prepare nutritious meals. Psychologically, many older people live alone and may not prepare food for one; financially, many live on social security and have a limited income.

Crossword Puzzle—

Across: ATROPHIC GASTRITIS, STRESS, STRESSORS, LONGEVITY, SARCOPENIA

Down: CONGREGATE MEALS, NEURONS, NEUROTRANSMITTER, EDENTULOUS, DYSPHAGIA

Short Answer—
1. See Table 17-3 in the text.

Sample Test Questions—
1. b (p. 588)
2. c (p. 589)
3. b (p. 589)
4. d (p. 589)
5. a (p. 592-593)
6. d (p. 593)
7. e (p. 593)
8. a (p. 594)
9. a (p. 595)
10. a (p. 595)
11. c (p. 597)
12. b (p. 598)
13. d (p. 598)
14. a (p. 599)
15. b (p. 602)

CHAPTER 18
DIET AND HEALTH

CHAPTER OUTLINE

I. Nutrition and Infectious Diseases
    A. The Immune System

    B. Nutrition and Immunity

    C. HIV and AIDS

    D. How AIDS Develops

    E. The HIV Wasting Syndrome

    F. Nutrition Support for People with HIV Infections

II. Nutrition and Chronic Diseases

III. Cardiovascular Disease
    A. How Atherosclerosis Develops

B. Risk Factors for Coronary Heart Disease

C. Recommendations for Reducing Coronary Heart Disease Risk

IV. Hypertension
    A. How Hypertension Develops

    B. Risk Factors for Hypertension

    C. Recommendations for Reducing Hypertension Risk

V. Diabetes Mellitus
    A. How Diabetes Develops

    B. Complications of Diabetes

    C. Recommendations for Diabetes

VI. Cancer
   A. How Cancer Develops
   _____
   _____
   _____
   _____

   B. Recommendations for Reducing Cancer Risk
   _____
   _____
   _____
   _____

VII. Recommendations for Chronic Diseases
   _____
   _____
   _____
   _____

Highlight: Complementary and Alternative Medicine
_____
_____
_____
_____
_____

SUMMING UP

Due to the control of (1)_____ diseases, the average life expectancy today is longer, and most diseases people now fear (other than AIDS) are: diseases of the (2)_____ and blood vessels, (3)_____, diabetes, lung diseases, and liver disease. Of the 10 leading causes of illness and death, 5 are directly associated with (4)_____. The risk factors for these degenerative diseases of adulthood are: (5)_____, behavioral, social, and (6)_____. (7)_____ is a risk factor which aggravates the risk of almost every other disease. Food (8)_____ are interwoven with these risk factors. Eating high (9)_____ foods and becoming (10)_____ increases the probabilities of contracting cancer, (11)_____, diabetes, (12)_____ and diverticulosis.

Lifestyle recommendations from the Surgeon General's Report on Nutrition and Health include: reduce consumption of fat and (13)_____, achieve and maintain a desirable body (14)_____, increase consumption of complex carbohydrates and (15)_____, reduce intake of (16)_____, use (17)_____ in moderation, if at all. The two types of diabetes mellitus are (18)_____-_____ and non-insulin dependent, which is most

219

common. The (19) _____ system was originally developed for people with diabetes but has been praised as the best possible system for people in perfect health to manage their diets.

Major causes of death in the U.S. have been from diseases of the heart and (20)_____ vessels. CVD accounts for more than (21)_____ of the nation's deaths per year, mostly by way of heart attacks and strokes. (22)_____, high blood pressure and high blood cholesterol are the 3 major risk factors for CVD.

The big diet-related risk factors for CVD are glucose intolerance, (23) _____, hypertension, and high blood cholesterol. Weight control, diet and (24) _____ are all key factors in controlling these adult related diseases.

A high dietary fat intake is thought to increase chances of developing (25) _____. Other factors related to cancer are: genetics, smoking, water and air pollution. Food (26) _____ probably have little to do with the causation of cancer. Dietary recommendations are to decrease intake of certain types of fats and increase plant (27) _____ intake. (28) _____ and tobacco use increase the risk of cancer. Nonnutrient compounds in the (29) _____ vegetables offer a protective effect against carcinogens.

Dietary (30) _____ (particularly excess food energy and fat intakes) increase the likelihood of diabetes, heart and blood vessel diseases, and cancer. With the exception of (31) _____ and drinking (32) _____, the (33) _____ choices can influence the long-term health prospects more than any other action a person can take.

(34)_____ compromises immunity. People with AIDS often experience malnutrition and (35)_____. Nutrition can prevent and reduce (36)_____.

Chapter Glossary

acquired immune deficiency syndrome (AIDS): the end stage of HIV infection, in which severe complications are manifested.
adenomas: cancers that arise from glandular tissues.
angina: a painful feeling of tightness or pressure in and around the heart, often radiating to the back, neck, and arms; caused by a lack of oxygen to an area of heart muscle.

antibodies: large proteins of the blood and body fluids, produced by the immune system in response to the invasion of the body by foreign molecules.
antigens: substances that elicit the formation of antibodies or an inflammation reaction from the immune system.
antipromoters: factors that oppose the development of cancer.
atherosclerosis: a condition characterized by plaques along the inner walls of the arteries.
autoimmune disorder: a condition in which the body develops antibodies to its own proteins and then proceeds to destroy cells containing these proteins.
B-cells: lymphocytes that produce antibodies.
benign: tumors that stop growing without intervention or can be removed surgically and pose no threat to health.
bioterrorism: the intentional spreading of disease-causing microorganisms or toxins.
cancers: diseases that result from the unchecked growth of malignant tumors.
carcinogens: substances or agents that are capable of causing cancer.
carcinomas: cancers that arise from epithelial tissues.
cardiovascular disease (CVD): a general term of all diseases of the heart and blood vessels.
CD4+ T-lymphocytes: circulating white blood cells that contain the CD4+ protein on their surfaces, and are a necessary component of the immune system.
CHD risk equivalents: disorders that raise the risk of heart attacks, strokes, and other complications associated with cardiovascular disease to the same degree as existing CHD.
coronary heart disease (CHD): the damage that occurs when the blood vessels carrying blood to the heart become narrow and occluded.
C-reactive protein (CRP): a protein produced during the acute phase of infection or inflammation that enhances immunity by promoting phagocytosis and activating platelets.
cytokines: special proteins that direct immune and inflammatory responses.
diabetes mellitus: a metabolic disorder of carbohydrate metabolism characterized by altered glucose regulation and utilization, usually caused by insufficient or relatively ineffective insulin.
embolism: the obstruction of a blood vessel by an embolus.
embolus: a traveling clot, causing sudden death.
emerging risk factors: recently identified factors that enhance the ability to predict disease risk in an individual.
fibrous plaques: mounds of lipid material, mixed with smooth muscle cells and calcium, which develop in the artery walls in atherosclerosis.
gangrene: the death of tissue, usually due to deficient blood supply.
genome: the full complement of genetic material (DNA) in the chromosomes of a cell.
gliomas: cancers that arise from glial cells of the central nervous system.
heart attack: sudden tissue death caused by blockages of vessels that feed the heart muscle; also called *myocardial infarction* or *cardiac arrest*.
Human Genome Project: an international project whose purpose is to determine the sequence of the human genome, develop genetic and physical maps of the human genome, locate and identify human genes, and explore the ethical, legal, and social implications of this work.
human immunodeficiency virus (HIV): the virus that causes AIDS.
hypertension: higher-than-normal blood pressure. Hypertension that develops without an identifiable cause is known as *essential* or *primary hypertension*; hypertension that is caused by a specific disorder such as kidney disease is known as *secondary hypertension*.
immune system: the body's natural defense system against foreign materials that have penetrated the skin or mucous membranes.
immunoglobulins: proteins capable of acting as antibodies.

impaired glucose tolerance: blood glucose levels higher than normal but not high enough to be diagnosed as diabetes; sometime called prediabetes.
initiation: an event caused by radiation or chemical reaction that can give rise to cancer.
initiators: factors that cause mutations that give rise to cancer.
insulin resistance: the condition in which a normal amount of insulin produces a subnormal effect; a metabolic consequence of obesity.
leukemias: cancers that arise from the white blood cells.
lymphocytes: white blood cells that participate in acquired immunity; B-cells and T-cells.
lymphomas: cancers that arise from lymph tissue.
malignant: tumors that multiply out of control, threaten health, and require treatment.
melanomas: cancers that arise from pigmented skin cells.
metabolic syndrome: a combination of four risk factors—insulin resistance, hypertension, abnormal blood cholesterol, and obesity—that greatly increase a person's risk of developing coronary heart disease.
metastasize: to spread from one part of the body to another.
microangiopathies: disorders of the small blood vessels.
neoplasm: a new growth of tissue forming an abnormal mass with no function.
opportunistic infections: infections from microorganisms that normally do not cause disease in the general population but can cause great harm in people once their immune systems are compromised (as in HIV infection).
peripheral resistance: resistance to the flow of blood caused by reduced diameter of the vessels at the periphery of the body—the smallest arteries and capillaries.
phagocytes: white blood cells that have the ability to ingest and destroy foreign substances.
phagocytosis: the process by which phagocytes engulf and destroy foreign materials.
plaques: mounds of lipid material, mixed with smooth muscle cells and calcium, that develop in the artery walls.
platelets: tiny, disc-shaped bodies in the blood, important in blood clot formation.
prehypertension: slightly higher-than-normal blood pressure, but not as high as hypertension.
promoters: factors that favor the development of cancer once it has started.
sarcomas: cancers that arise from muscle, bone, or connective tissues.
stroke: an event in which the blood flow to a part of the brain is cut off; also called *cerebrovascular accident (CVA)*.
synergistic: multiple factors operating together in such a way that their combined effects are greater than the sum of their individual effects.
thrombosis: the formation of a thrombus.
thrombus: a blood clot that may obstruct a blood vessel, causing gradual tissue death.
transient ischemic attack (TIA): a temporary reduction in blood flow to the brain, which causes temporary symptoms that vary depending of the part of the brain affected.
T-cells: lymphocytes that attack antigens.
thrombosis: the formation or development of a thrombus.
thromboxanes: eicosanoid compounds with effects on the blood-clotting system.
thrombus: a blood clot that may obstruct a blood vessel or the heart cavity.
tumor: a new growth of tissue forming an abnormal mass with no function; also called a *neoplasm*. Tumors that pose no problem are called benign; those that resist treatment and are harmful are *malignant*.
type 1 diabetes: the less common type of diabetes in which the person produces no insulin at all.
type 2 diabetes: the more common type of diabetes in which the fat cells resist insulin.
wasting syndrome: an involuntary loss of more than 10% of body weight, common in AIDS and cancer.

Assignments

Answer these chapter study questions from the text:

1. How do the major diseases of today as a group differ from those of several decades ago as a group? Why is nutrition considered so important in connection with today's major diseases?

2. What is HIV infection? What are the consequences of HIV infection? What is the HIV wasting syndrome?

3. In what ways might good nutrition status possibly alter the course of HIV infection?

4. Identify the major diet-related risk factors for atherosclerosis, hypertension, cancer, and diabetes.

5. Describe some ways in which people can alter their diets to lower their blood cholesterol levels.

6. Describe some steps that people with hypertension can take to lower their blood pressure.

7. Name the two major types of diabetes and describe some differences between them. How do dietary recommendations for each type of diabetes compare with the healthy diet recommended for all people?

8. Differentiate between cancer initiators, promoters, and antipromoters. Which nutrients or foods fit into each of these categories?

9. Describe the characteristics of a diet that might offer the best protection against the onset of cancer.

Complete these short answer questions:

1. The 4 steps in cancer development are:

    a.

    b.

    c.

    d.

2. List 3 ways a high-fat diet may promote cancer:

    a.

    b.

    c.

Complete this crossword puzzle by Mary A. Wyandt, Ph.D., CHES.

Across:	Down:
2. tumors that multiply out of control, threaten health, and require treatment	1. tumors that stop growing without intervention or can be removed surgically and pose no threat to health
4. substances or agents that are capable of causing cancer	3. cancers that arise from lymph tissue
8. a new growth of tissue forming an abnormal mass with no function	5. cancers that arise from glandular tissue
9. another word for neoplasm; those that pose no problem are called benign; those that resist treatment and are harmful are malignant	6. the obstruction of a blood vessel by an embolus
10. to spread from one part of the body to another	7. cancers that arise from glial cells of the central nervous system

SAMPLE TEST QUESTIONS

1. The diseases most feared today are:

    1. tuberculosis
    2. smallpox
    3. diseases of the heart and blood vessels
    4. cancer
    5. diabetes

    a. 1, 2, 3   c. 3, 4, 5
    b. 2, 3, 4   d. 1, 3, 5

2. This disease has some relationship with nutrition:

    a. cancer.   d. a and b
    b. diabetes.   e. a, b, and c
    c. heart disease.

3. Being obese increases the probabilities of contracting which of the following?

    a. cancer   d. atherosclerosis
    b. hypertension   e. all of the above
    c. diabetes

4. Mounds of lipid material mixed up with smooth muscle cells and calcium which develop in the artery walls are called:

    a. osteoporosis.   c. CVD.
    b. atherosclerosis.   d. fibrous plaques.

5. The event in which an embolus lodges in vessels that feed the heart muscle causing sudden tissue death is called a:

    a. thrombus.   c. stroke.
    b. heart attack.   d. aorta.

6. Damage that occurs when the blood vessels carrying blood to the heart become narrow and occluded is called:

    a. an aneurysm.   c. an aorta.
    b. a plaque.   d. coronary heart disease.

7. These two conditions worsen each other:

    a. cancer, diverticulosis.   c. hypertension, atherosclerosis.
    b. cancer, diabetes.   d. hypoglycemia, hypertension.

8. Risk factors for atherosclerosis that can be minimized by behavior change include:

   1. smoking
   2. hypertension
   3. gender
   4. lack of exercise
   5. obesity
   6. heredity
   7. stress

   a. 1,2,3,4,5
   b. 1,2,4,5,6
   c. 2,3,4,5,7
   d. 1,2,4,5,7

9. A high-fat, high-cholesterol diet can increase heart disease risk.

   a. True
   b. False

10. The first action to take to lower blood cholesterol should be:

    a. achieve and maintain appropriate body weight.
    b. consume large amounts of fish and fish oils.
    c. begin drug treatment.
    d. consume a high protein diet.

11. Many people with hypertension are medically advised to lower their salt intakes.

    a. True
    b. False

12. Diseases that result from the unchecked growth of malignant tumors are:

    a. AIDS.
    b. cancers.
    c. hypertension.
    d. diverticulosis.

13. This type of diet is linked to cancer development:

    a. high-fat.
    b. low-fat.
    c. high-fiber.
    d. high food additive.

14. In this condition, the pancreas is unable to synthesize the hormone insulin.

    a. AIDS
    b. Type II diabetes mellitus
    c. cancer
    d. Type I diabetes mellitus

15. This condition is characterized by insulin resistance of the fat cells.

    a. Type I diabetes mellitus
    b. Type II diabetes mellitus
    c. hypoglycemia
    d. atherosclerosis

16. The death of tissue, usually due to deficient blood supply is:

    a. reactive hypoglycemia.
    b. non-reactive hypoglycemia.
    c. insulin resistance.
    d. gangrene
    e. glucosuria.

17. The person with diabetes is at a high risk of developing:

    a. cancer.
    b. diverticulosis.
    c. AIDS.
    d. strokes and heart attacks.

18. The causes of malnutrition and wasting in HIV infection:

    a. are related to the disease itself.
    b. are related to the treatments for the disease.
    c. are related to anorexia.
    d. a and b
    e. a, b and c

19. Dietary treatments for HIV wasting are aimed primarily at:

    a. providing adequate vitamins and minerals.
    b. meeting nutrient needs.
    c. reducing malnutrition.
    d. providing nutrient-dense products.
    e. all of the above

ANSWERS

Summing Up—(1) infectious; (2) heart; (3) cancer; (4) nutrition; (5) environmental; (6) genetic; (7) obesity; (8) behaviors; (9) fat; (10) obese; (11) hypertension; (12) atherosclerosis; (13) cholesterol; (14) weight; (15) fiber; (16) sodium; (17) alcohol; (18) insulin-dependent; (19) exchange; (20) blood; (21) half; (22) smoking; (23) obesity; (24) exercise; (25) cancer; (26) additives; (27) fiber; (28) alcohol; (29) cruciferous; (30) excesses; (31) smoking; (32) alcohol; (33) dietary; (34) Malnutrition; (35) wasting; (36) malnutrition.

Chapter study questions from the text—(1) The major diseases of today are related to lifestyle, whereas the diseases of several decades ago were mostly infectious. Nutrition is considered so important in connection with today's major diseases because it is involved with risk factors. (2) HIV infection is human immunodeficiency virus; it attacks the immune system and causes those who are infected to be vulnerable to opportunistic infections. It is not curable and in its late stages it usually causes severe weight loss, tuberculosis, recurrent bacterial pneumonia, CNS infections, GI tract infections, skin infections, cancers, and severe diarrhea. HIV wasting syndrome is the malnutrition and wasting involved with the disease; it is an involuntary loss of more than 10% of body weight. (3) Attention to nutrition may prevent and reverse malnutrition, which may improve the quality of life and slow disease progression. (4) Risk factors are environmental, behavioral, social and genetic; specific risk factors are: high-fat diet, excessive alcohol consumption, high salt intake, contaminated food intake, low complex carbohydrate and fiber intake, low vitamin and/or mineral intakes, high sugar intake, low calcium intake, stress, sedentary lifestyle, smoking, obesity. (5) Maintain appropriate body weight, exercise regularly, do not smoke, control stress, control blood cholesterol, consume a low-fat diet. (6) Balance energy intake and output to maintain appropriate body weight, lower salt intake, exercise, eat adequate amounts of calcium, lower saturated fat intake, and, if you drink alcohol, do so in moderation. (7) Two types of diabetes are insulin-dependent diabetes mellitus (IDDM) or type I and non-insulin dependent diabetes mellitus (NIDDM) or type II. In type I the person produces no insulin; type II is more common and usually arises in adulthood. Exchange lists are used to plan diets, and general dietary guidelines for good health are similar to meal planning for diabetes. (8) Initiators are carcinogens that intrude into cells and alter the genetic material. Promoters enhance cancer development; antipromoters offer protective effects. Additives, pesticides, nitrosamines, and alcohol are initiators; fat and linoleic acid are promoters; fruits, vegetables, fiber and antioxidants are antipromoters. (9) Avoid obesity, reduce the consumption of total fat, eat more high-fiber foods, such as whole-grain cereals, vegetables, and fresh fruits, include a variety of vegetables and fruits in the daily diet, avoid possible carcinogens by limiting consumption of foods preserved by salt, smoke, or nitrites, limit or stop consumption of alcoholic beverages.

Short Answers—

1.
   a. exposure to a carcinogen.
   b. entry of the carcinogen into a cell.
   c. initiation, probably by the carcinogen's altering the cellular DNA.
   d. enhancement of cancer development by promoters, probably involving several more steps before the cell begins to multiply out of control; tumor formation.
2.
   a. by causing the body to secrete more of certain hormones (for example, estrogen), thus creating a climate favorable to the development of certain cancers (for example, breast cancer).
   b. by promoting the secretion of bile into the intestine; bile may then be converted by organisms in the colon into compounds that cause cancer.
   c. by being incorporated into cell membranes and changing them so that they offer less defense against cancer-causing invaders.

Crossword Puzzle—

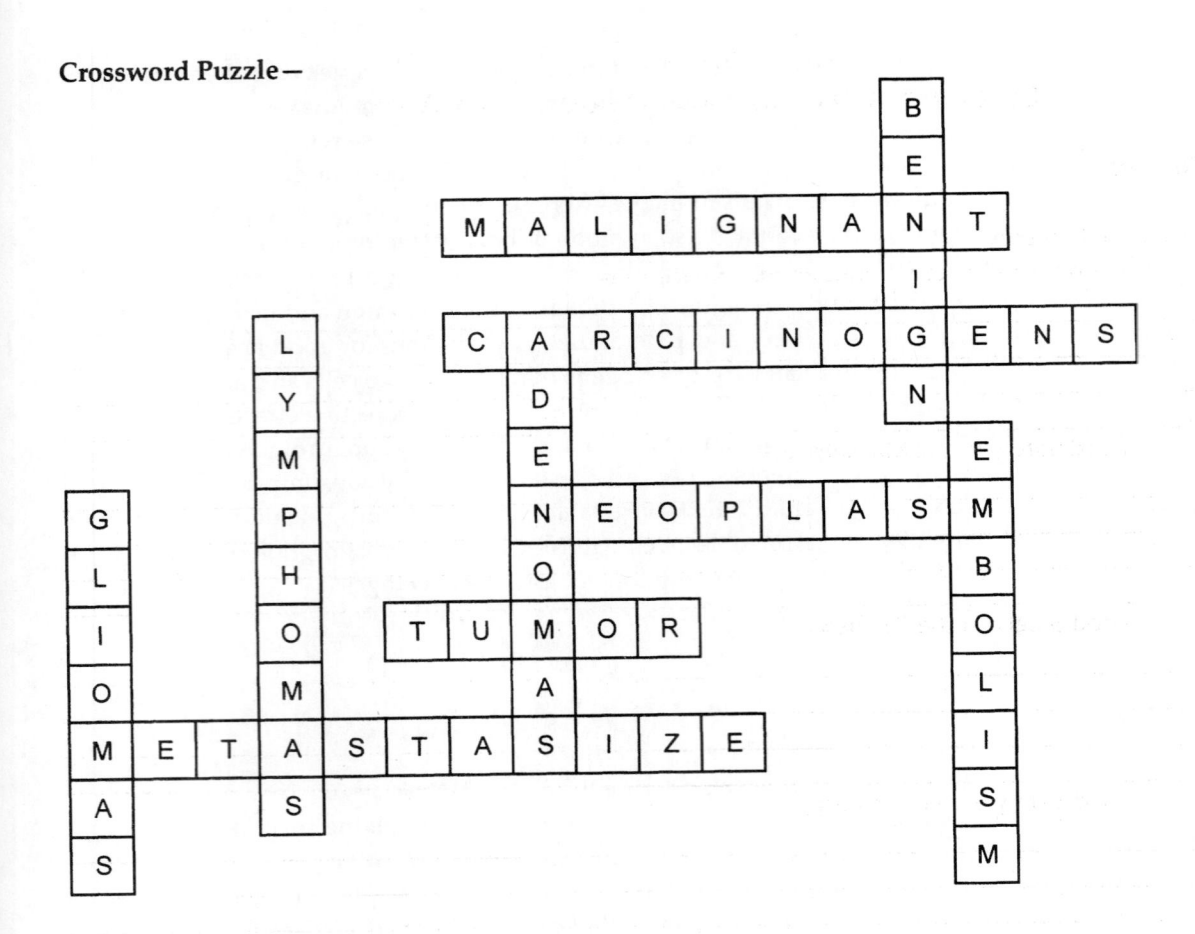

```
                                B
                                E
        M  A  L  I  G  N  A  N  T
                                I
     L           C  A  R  C  I  N  O  G  E  N  S
     Y              D                 N
     M              E                    E
  G  P           N  E  O  P  L  A  S  M
  L  H              O                    B
  I  O        T  U  M  O  R              O
  O  M              A                    L
  M  E  T  A  S  T  A  S  I  Z  E        I
  A     S                                S
  S                                      M
```

Sample Test Questions—

1. c (p. 613)
2. e (p. 614)
3. e (p. 614)
4. d (p. 620)
5. b (p. 621)
6. d (p. 620)
7. c (p. 621)
8. d (p. 621)
9. a (p. 622)
10. a (p. 623)
11. a (p. 627)
12. b (p. 636)
13. a (p. 639)
14. d (p. 632)
15. b (p. 633)
16. d (p. 634)
17. d (p. 634)
18. e (p. 617)
19. e (p. 617-618)

Chapter 19
Consumer Concerns about Foods and Water

Chapter Outline

I. Foodborne Illnesses
   A. Foodborne Infections and Food Intoxications

   B. Food Safety in the Marketplace

   C. Food Safety in the Kitchen

   D. Food Safety While Traveling

   E. Advances in Food Safety

II. Nutritional Adequacy of Foods and Diets
   A. Obtaining Nutrient Information

   B. Minimizing Nutrient Losses

III. Environmental Contaminants
   A. Harmfulness of Environmental Contaminants

    B.   Guidelines for Consumers

IV.   Natural Toxicants in Foods

V.   Pesticides
    A.   Hazards and Regulation of Pesticides

    B.   Monitoring Pesticides

    C.   Consumer Concerns

VI.   Food Additives
    A.   Regulations Governing Additives

    B.   Intentional Food Additives

    C.   Indirect Food Additives

VII.   Consumer Concerns about Water
    A.   Sources of Drinking Water

B. Water Systems and Regulations

Highlight: Food Biotechnology

SUMMING UP

Foods contain other substances besides nutrients: (1)_____, pesticides, and additives. Foodborne illnesses refer to food borne (2)_____ and food intoxicants and they top the FDA's list of food hazard priorities. Foodborne infections are cause by eating food contaminated by infectious (3)_____, food intoxicants are caused by eating foods containing (4)_____ toxins or microbes that produce toxins.

A contaminant of a food is anything that does not belong there including (5)_____, harmful substances from industry, (6)_____ residues, bits of packaging and ordinary dirt. Many commonly used plants and plant products contain naturally occurring (7)_____ substances. Pesticides leave (8)_____ in foods; their use is regulated so as not to present a hazard to consumers. Consumer groups are not reassured by the regulation of pesticides and encourage (9)_____ farming—a process impractical for the entire population.

Food (10)_____ reduce the risk of foodborne illness and prevent food spoilage. The (11)_____ regulates the use of food additives and maintains a GRAS (generally recognized as safe) list. Intentional food additives include: artificial (12)_____, artificial flavors and flavor enhancers, (13)_____ agents, antioxidants, (14)_____ additives, and radiation. Incidental food additives find their way into food as the result of harvesting, production, processing, storage or packaging. The nutritional (15)_____ of foods ranks high among the FDA's issues. The FDA has become increasingly involved in the proper (16)_____ and accurate advertising of products.

Chapter Glossary

additives: substances not normally consumed as foods but added to food either intentionally or by accident.
antimicrobial agents: preservatives that prevent microorganisms from growing.
antioxidants: preservatives that prevent rancidity of fats in foods and other damage to food caused by oxygen.
artesian water: water that is drawn from a well that taps a confined aquifer in which the water level stands above the natural water table.
artificial colors: certified (approved by the FDA) food colors added to enhance appearance.
artificial flavors, flavor enhancers: chemicals that mimic natural flavors and those that enhance flavor.
BHA and BHT: preservatives commonly used to slow the development of off-flavors, odors, and color changes caused by oxidation.
bioaccumulation: the accumulation of contaminants in the flesh of animals high on the food chain.
bovine growth hormone (BGH): a hormone produced naturally in the pituitary gland of a cow that promotes growth and milk production; now produced for agricultural use by transgenic bacteria.
certification: the process in which a private laboratory inspects shipments of a product for selected chemicals and then, if the product is free of violative levels of those chemicals, issues a guarantee to that effect.
contaminants: substances that make food impure and unsuitable for ingestion.
cross-contamination: the contamination of cooked food by bacteria that occurs when the food comes in contact with surfaces touches by raw meat, poultry, or seafood.
Delaney Clause: a clause in the Food Additive Amendment to the Food, Drug, and Cosmetic Act that states that no substance that is known to cause cancer in animals or human beings at any dose level shall be added to foods.
dioxins: a class of chemical pollutants created as by-products of chemical manufacturing, incineration, chlorine bleaching of paper pulp, and other industrial processes.
distilled water: water that has been vaporized and recondensed, leaving it free of dissolved minerals.
emulsifiers and gums: thickening and stabilizing agents that maintain emulsions, foams, or suspensions.
filtered water: water treated by filtration, usually through activated carbon filters that reduce the lead in tap water, or by reverse osmosis units that force pressurized water across a membrane removing lead, arsenic, and some microorganisms.
fluoridated water: water that has been treated so as to contain at least 0.8 milligrams of fluoride per liter.
foodborne illness: illness transmitted to human beings through food, caused by either an infectious agent (*food-borne infection*) or a poisonous substance (*food intoxication*); commonly known as *food poisoning*.
food chain: the sequence in which living things depend on other living things for food.
GRAS (generally recognized as safe): food additives that have long been in use and are believed safe.
hard water: water with a high calcium and magnesium concentration.
hazard: source of danger; used to refer to circumstances in which toxicity is possible under normal conditions of use.
Hazard Analysis Critical Control Points (HACCP): a systematic plan to identify and correct potential microbial hazards in the manufacturing, distribution, and commercial use of food products.
heavy metal: any of a number of mineral ions such as mercury and lead, so called because they are of relatively high atomic weight. Many heavy metals are poisonous.
indirect or incidental additives: substances that can get into food as a result of contact with foods during growing, processing, packaging, storing, cooking, or some other stage before the foods are consumed; also called *accidental additives*.

intentional food additives: additives intentionally added to foods, such as nutrients, colors, and preservatives.
irradiation: sterilizing a food by exposure to energy waves, similar to ultraviolet light and microwaves.
margin of safety: when speaking of food additives, a zone between the concentration normally used and that at which a hazard exists.
mineral water: water from a spring or well that typically contains 250 to 400 ppm of minerals.
monosodium glutamate (MSG): a sodium salt of the amino acid glutamic acid commonly used as a flavor enhancer.
MSG symptom complex: an acute, temporary intolerance reaction that may occur after the ingestion of the additive MSG.
natural water: water obtained from a spring or well that is certified to be safe and sanitary.
nitrates: salts that are converted to nitrites by bacteria.
nitrites: salts added to food to prevent botulism; one example is sodium nitrite, which is used to preserve meats.
nitrosamines: derivatives of nitrites that may be formed in the stomach when nitrites combine with amines; nitrosamines are carcinogenic in animals.
nutrient additives: vitamins and minerals added to improve nutritive value.
organic halogen: an organic compound containing one or more atoms of a halogen—fluorine, chlorine, iodine, or bromine.
organic: in agriculture, crops grown and processed according to USDA regulations defining the use of fertilizers, herbicides, insecticides, fungicides, preservatives and other chemical ingredients.
pasteurization: heat processing of food that inactivates some, but not all, microorganisms in the food; not a sterilization process.
pathogens: microorganisms or substances capable of producing disease.
PBB (polybrominated biphenyl) and PCB (polychlorinated biphenyl): toxic organic compounds used in pesticides, paints, and flame retardants.
persistence: stubborn or enduring continuance; with respect to food contaminants, the quality of persisting, rather than breaking down, in the bodies of animals and human beings.
pesticides: chemicals used to control insects, diseases, weeds, fungi, and other pests on plants, vegetables, fruits, and animals. Used broadly, the term includes herbicides (to kill weeds), insecticides (to kill insects), and fungicides (to kill fungi).
potable water: water that is suitable for drinking.
preservatives: antimicrobial agents, antioxidants, and other additives that retard spoilage or maintain desired qualities, such as softness in baked goods.
public water: water from a municipal or county water system that has been treated and disinfected.
purified water: water that has been processed through distillation, deionization, or reverse osmosis and meets U.S. Pharmacopoeia standards for medical and research purposes.
radiolytic products: chemicals formed during the irradiation of food.
residues: whatever remains. In the case of pesticides, those amounts that remain on foods when people buy and use them.
risk: a measure of the probability and severity of doing harm.
safety: the condition of being free from harm or danger.
soft water: water with a high sodium concentration.
solanine: a poisonous narcotic-like substance present in potato peels and sprouts.
spring water: water originating from an underground spring or well.
sulfites: salts containing sulfur, added to foods to prevent spoilage.
sushi: vinegar-flavored rice and seafood, typically wrapped in seaweed and stuffed with colorful vegetables.

thickening and stabilizing agents: ingredients that maintain emulsions, foams, or suspensions or lend a desirable thick consistency to foods.

tolerance level: the maximum amount of a residue permitted in a food when a pesticide is used according to label directions.

toxicity: the ability of a substance to harm living organisms. All substances are toxic if high enough concentrations are used.

traveler's diarrhea: nausea, vomiting, and diarrhea caused by consuming food or water contaminated by any of several organisms.

ultrahigh temperature (UHT) treatment: short-time exposure of a food to temperatures above those normally used in order to sterilize it.

well water: water drawn from groundwater by tapping into an aquifer.

ASSIGNMENTS

Answer these chapter study questions from the text:

1. To what extent does food poisoning present a real hazard to consumers eating U.S. foods? How often does it occur?

2. Distinguish between the two types of foodborne illnesses and provide an example of each. Describe several measures that help prevent foodborne illnesses.

3. What special precautions apply to meats? To seafood?

4. What is meant by a "persistent" contaminant of foods? Describe how contaminants get into foods and build up in the food chain.

5. What dangers do natural toxicants present?

6. How do pesticides become a hazard to the food supply, and how are they monitored? In what ways can people reduce the concentrations of pesticides in and on foods that they prepare?

7. What is the difference between a GRAS substance and a regulated food additive? Give examples of each. Name and describe the different classes of additives.

Complete this crossword puzzle by Mary A. Wyandt, Ph.D., CHES.

	Across:		Down:
2.	the process in which a private laboratory inspects shipments of a product for selected chemicals and then, if the product is free of volatile levels of those chemicals, issues a guarantee to that effect	1.	stubborn or enduring continuance
4.	a poisonous narcotic-like substance present in potato peels and sprouts	3.	the maximum amount of a residue permitted in a food when a pesticide is used according to label directions
5.	sterilizing a food by exposure to energy waves, similar to ultraviolet light and microwaves	6.	the condition of being free from harm or danger
7.	microorganisms or substances capable of producing disease	8.	source of danger; used to refer to circumstances in which toxicity is possible under normal conditions of use
10.	whatever remains; in the case of pesticides, those amounts that remain on foods when people buy and use them	9.	a measure of the probability and severity of doing harm

Sample Test Questions

1. First on the FDA's list of priority concerns related to food is:

    a. foodborne infection.
    b. environmental contaminants.
    c. pesticide residues.
    d. food additives.
    e. food toxicants.

2. Food additives are the concern of the:

    a. GRAS.
    b. USDA.
    c. FDA.
    d. CFC.

3. Substances widely used for many years without apparent ill effects are on this list.

    a. FDA
    b. GRAS
    c. Delaney
    d. Additive Safety

4. The Delaney Clause states that no additives known to cause cancer in:

    a. animals or people at any dose level can be used.
    b. people at any dose level can be used.
    c. animals or people at one gram levels can be used.
    d. people at one gram levels can be used.

5. Food additives may be used to:

    a. disguise faulty products.
    b. deceive customers.
    c. destroy nutrients.
    d. enhance flavor.

6. One of the best known flavoring agents is:

    a. tartrazine.
    b. UHP.
    c. monosodium glutamate.
    d. phenylalanine.

7. Antimicrobial agents protect foods against:

    a. oxidation.
    b. organisms.
    c. radiation.
    d. pesticides.

8. Which of the following are *not* added to foods to prevent oxidation?

    a. vitamins C and E
    b. sulfites
    c. BHA and BHT
    d. MSG and nitrites

9. Bottled water is superior to the public water supply because it is always free from contaminants.

   a. True   b. False

10. Incidental additives find their way into food as a result of:

    a. increasing nutritive value of foods.
    b. home cooking errors.
    c. manufacturing procedures.
    d. advertising gimmicks.

11. Pesticide use is not monitored by:

    a. USPS.   d. EPA.
    b. FAO.   e. FDA.
    c. WHO.

12. When cooked foods come in contact with surfaces touched by raw meat, this is called:

    a. HACCP.   c. irradiation.
    b. cross contamination. d. irradiation

13. The major source of contaminants, such as heavy metals, in the food chain is:

    a. industry.   c. nature.
    b. agriculture. d. microorganisms.

14. Food poisoning is caused by food contaminated by:

    a. heavy metals. c. goitrogens.
    b. antioxidants. d. microorganisms.

15. The USDA seal on raw meat and poultry guarantees protection from bacterial contamination.

    a. True   b. False

16. The safest way to thaw meats or poultry is to leave them on the counter at room temperature.

    a. True   b. False

17. An example of an organism that may be present in anaerobic conditions is:

    a. Salmonella.
    b. Botulinum.
    c. hepatitis.
    d. nitrosamine.
    e. Giardia.

Answers

Summing Up—(1) contaminant; (2) infections; (3) microbes; (4) natural; (5) microbes; (6) pesticide; (7) toxic; (8) residues; (9) organic; (10) additives; (11) FDA; (12) colors; (13) antimicrobial; (14) nutrient; (15) adequacy; (16) labeling.

Chapter study questions from the text—(1) Foodborne illness is the leading food safety concern. 6.5 million cases are reported, and millions more go unreported. (2) Two types of foodborne illnesses include those caused by an infectious agent (foodborne infection), example—*Campylobacter jejuni*; and those caused by a poisonous substance (food intoxication), example—*Clostridium botulinum*. Measures to prevent them include: use proper canning methods, avoid commercially canned foods with leaky seals, or with bent, bulging or broken cans, cook food thoroughly, use sanitary food handling methods, avoid unpasteurized milk, avoid raw fruits and vegetables where protozoa are endemic, dispose of sewage properly, refrigerate foods promptly and properly. (3) Wash all surfaces that have been in contact with raw meats, poultry, eggs, fish and shellfish before reusing; serve cooked meats, poultry and seafood on a clean plate. Separate raw meats and seafood from those that have been cooked. Do not use marinade that was in contact with raw meat. When cooking meats, use a thermometer to test the internal temperature, and cook to the temperature indicated for that particular meat. Cook hamburgers to at least medium well-done. Cook stuffing separately or stuff poultry just prior to cooking. Do not cook large cuts of meat or turkey in a microwave oven. Cook eggs before eating them. Cook seafood thoroughly. When serving meats and seafood, maintain temperature of 140 degrees or higher, and heat leftovers thoroughly to least 165 degrees. (4) Stubborn or enduring continuance; the quality of persisting, rather than breaking down, in the bodies of animals and human beings. Contaminants get into foods: heavy metals and other contaminants entering the air in smokestack emissions return to the soil in rainfall, contaminants in the soil are absorbed by plants. People either eat the plants (fruits and vegetables) or meat from livestock that have eaten the plants. Sewage sludge and pesticides leave residues in the soil; runoff pollutes ground and surface water and contaminates the seafood that people eat. Toxins in the food chain accumulate: a person whose principal animal-protein source is fish may consume about 100 pounds of fish in a year, and these fish will have consumed a few tons of plant-eating fish in the course of their lifetimes; the plant eaters will have consumed several tons of photosynthetic producer organisms. If the producer organisms have become contaminated with toxic chemicals, these chemicals become more concentrated in the bodies of the fish that consume them. If none of the chemicals are lost along the way, one person ultimately eats the same amount of contaminant as was present in the original several tons of producer organisms. (5) Poisonous mushrooms are natural yet can be dangerous when eaten. Cabbage, turnips, mustard greens, and radishes contain small quantities of goitrogens that can enlarge the thyroid gland; this can cause problems if a person with a thyroid problem consumes large quantities of these foods. Lima beans and some fruit seeds contain cyanogens that, if activated, can produce the deadly poison cyanide. Potatoes contain small amounts of natural poisons. Poisons are poisons whether made by man or by nature. (6) By remaining on crops, polluting water, contaminating the soil, and accumulating in the tissues of animals. Not well monitored but under the watch of FAO, WHO, and EPA. Buy fresh foods grown locally, using responsible methods, and buy a variety of foods. (7) A GRAS substance (such as salt) is accepted as safe based on long experience consistent with the belief that it is not hazardous, whereas a regulated food additive (such as MSG) has been chemically tested to ensure its effectiveness and safety.

Crossword Puzzle—

```
        P
    C E R T I F I C A T I O N
        R       O
        S   S O L A N I N E
        I       E
        S       R
        T   I R R A D I A T I O N
        E       N
        N   S   C
        C   P A T H O G E N S   R
        E   F   A       L       I
            E   Z   R E S I D U E S
            T   A   V           K
            Y   R   E
                D   L
```

Sample Test Questions--

1.	a (p. 658)	7.	b (p. 677)	13.	a (p. 661)
2.	c (p. 675)	8.	d (p. 677-678)	14.	d (p. 659)
3.	b (p. 675)	9.	b (p. 682)	15.	b (p. 663)
4.	a (p. 675)	10.	c (p. 679)	16.	b (p. 663)
5.	d (p. 675)	11.	a (p. 658)	17.	b (p. 659)
6.	c (p. 678)	12.	b (p. 661)		

Chapter 20
Hunger and the Global Environment

Chapter Outline

I. Hunger in the United States
   A. Defining Hunger in the United States

   B. Relieving Hunger in the United States

II. World Hunger
   A. Food Shortages

   B. Malnutrition

   C. Diminishing Food Supply

III. Poverty and Overpopulation

IV. Environmental Degradation and Hunger
   A. Environmental Limitations in Food Production

   B. Other Limitations in Food Production

V. Solutions
   A. Sustainable Development Worldwide
   _____
   _____
   _____
   _____

   B. Activism and Simpler Lifestyles at Home
   _____
   _____
   _____

Highlight: Progress Toward Sustainable Food Production
_____
_____
_____
_____

SUMMING UP

Consumers grow curious about the (1)_____ impacts of their food choices. Alternative food choices that are more environmentally benign are now (2)_____. This chapter emphasizes personal lifestyle (3)_____ because they raise awareness of the need for larger actions and (4)_____ the way for them.

Many trends are occurring that contribute to global (5)_____ that are related in that their (6)_____ overlap as well as their (7)_____. Environmentally conscious food (8)_____, preparing, cooking, and transporting can assist in solving related problems. Using proper cooking (9)_____ and kitchen appliances can also be beneficial.

This chapter also addresses the problems of (10)_____, and poverty in the United States and (11)_____ countries. These problems have always existed and despite numerous (12)_____, the number of hungry and poor people continues to grow.

Many nations now recognize that improvement of all nations' economies is a prerequisite to meeting the world's other urgent needs: (13)_____ stabilization, arrest of environmental (14)_____, sustainable treatment of (15)_____, and relief of hunger.

Chapter Glossary

carrying capacity: the total number of living organisms that a given environment can support without deteriorating in quality.

cash crops: crops grown for cash, as opposed to crops grown for food; examples include cotton and tobacco.

emergency kitchen: programs that provide groceries to be eaten on site; often called soup kitchens.

famine: extreme scarcity of food in an area that causes starvation and death in a large portion of the population.

food bank: a central source for food donation and distribution to local charities feeding the hungry.

food insecurity: limited or uncertain access to foods of sufficient quality and quantity to sustain a healthy and active live.

food insufficiency: an inadequate amount of food due to a lack of resources.

food pantries: programs that provide groceries to be prepared and eaten at home.

food poverty: hunger occurring from inadequate access to available food.

food recovery: collecting wholesome food for distribution to low-income people who are hungry; four common methods include *field gleaning, perishable food rescue or salvage, prepared food rescue, and nonperishable food collection.*

food security: certain access to enough food for all people at all times to sustain a healthy and active life.

fossil fuels: coal, oil, and natural gas; these are nonrenewable fuels that pollute. (Renewable or alternative fuels, such as solar and wind energy, pollute less or not at all.)

human carrying capacity: the maximum number of people the earth can support over time.

oral rehydration therapy (ORT): the administration of a simple solution of sugar, salt and water, taken by mouth, to treat dehydration caused by diarrhea.

sustainable: able to continue indefinitely. Here, the term means the use of resources at such a rate that the earth can keep on replacing them; for example, cutting trees no faster than new ones grow and producing pollutants at a rate with which the environment and human cleanup efforts can keep pace, so that no net accumulation of pollution occurs.

Assignments

Answer these chapter study questions from the text:

1. Identify some reasons why hunger is present in a country as wealthy as the United States.

2. Identify some reasons why hunger is present in the developing countries of the world.

3. Explain why relieving environmental problems will also help to alleviate hunger and poverty.

4. Discuss the different paths by which rich and poor countries can attack the problems of world hunger and the environment.

5. Describe some strategies that consumers can use to minimize negative environmental impacts when shopping for food, preparing meals, and disposing of garbage.

SAMPLE TEST QUESTIONS

1. An estimated one out of every _____ people worldwide live in poverty:

    a. 4
    b. 6
    c. 8
    d. 10
    e. 20

2. Collecting crops from fields that have already been harvested is:

    a. food insecurity.
    b. field gleaning.
    c. food rescue.
    d. food collection.

3. Extreme scarcity of food that causes starvation is:

    a. extinction.
    b. deforestation.
    c. sustainable.
    d. famine.
    e. poverty.

4. When selecting foods, the most constructive guideline is to buy more:

    a. vegetables, fruits, and grains.
    b. meat products.
    c. dairy products.
    d. processed food products.

5. To help prevent measles mortality and blindness, vitamin A supplements are distributed worldwide.

   a. True   b. False

6. Administration of a sugar, salt and water solution is called:

   a. ORT.   c. WHO.
   b. FDA.   d. ADA.

7. Using resources at a rate that the earth can continue replacing them is called.

   a. degradation.   c. population growth.
   b. sustainable.   d. carrying capacity.

8. The primary cause of hunger is:

   a. lack of education about food selection and preparation.
   b. abuse of alcohol and other drugs.
   c. depression.
   d. poverty.
   e. mental illness.

9. Communities can address their hunger problems by:

   a. improving public transportation to food resources.
   b. integrating public and private resources to relieve hunger.
   c. identifying high-risk populations and target services to meet their needs.
   d. all of the above

10. Ground-level ozone pollution and outer-atmosphere ozone depletion reduce agricultural outputs.

    a. True   b. False

11. Overpopulation and environmental degradation worsen poverty.

    a. True   b. False

12. Factors that affect population growth include:

    a. birth rates.
    b. death rates.
    c. standards of living.
    d. morbidity rates.
    e. a, b, and c

Complete this crossword puzzle by Mary A. Wyandt, Ph.D., CHES.

Across:	Down:
1. a common method of food recovery; _____ food rescue	2. programs that provide groceries to be eaten on site; often called soup kitchens
4. coal, oil, and natural gas; these are nonrenewable fuels that pollute	3. a common method of food recovery known as _____ or *perishable food rescue*
5. _____ or *alternative fuels*, such as solar and wind energy, pollute less or not at all	4. a common method of food recovery known as _____ *gleaning*
7. an inadequate amount of food due to a lack of resources	6. the maximum number of people the earth can support over time is called the *human* _____ *capacity*
9. a white, fluffy cash crop	8. (abbrv.) the administration of a simple solution of sugar, salt and water, taken by mouth, to treat dehydration caused by diarrhea

ANSWERS

Summing Up—(1) environmental; (2) available; (3) choices; (4) pave; (5) problems; (6) causes; (7) solutions; (8) shopping; (9) methods; (10) hunger; (11) developing; (12) programs; (13) population; (14) degradation; (15) resources.

Chapter study questions from the text—(1) Hunger is present in the U.S. because of the many political, social, and economic factors related to poverty. (2) Poverty, lack of clean drinking water, illiteracy, famine, and diminishing food supply. (3) Relieving environmental problems can help eliminate hunger and poverty in several ways: by lowering soil erosion, the agricultural productivity will improve; decreasing deforestation will decrease droughts and floods; lowering air pollution will increase crop yields. (4) The poor nations need to reduce population growth; reverse destruction of forests, waterways, and soil; and reduce poverty. Rich nations need to reduce wasteful and polluting uses of resources and energy; and relieve debtor nations of poverty. (5) Shoppers can car pool, use mass transit, walk, or bicycle to shop, shop less often—make fewer trips, or take turns shopping for each other. When preparing foods: cook mostly plant foods, use a pressure cooker and microwave or stir-fry foods, use oven and stove top less often, do without small electrical appliances. When disposing garbage: find uses for items normally thrown away, recycle trash, refuse to purchase items that have excess packaging.

Sample Test Questions--

1.	c (p. 693)	5.	a (p. 698)	9.	d (p. 696)		
2.	b (p. 696)	6.	a (p. 699)	10.	a (p. 702)		
3.	d (p. 697)	7.	b (p. 703)	11.	a (p. 700)		
4.	a (p. 705)	8.	d (p. 696)	12.	e (p. 699-701)		

Crossword Puzzle—

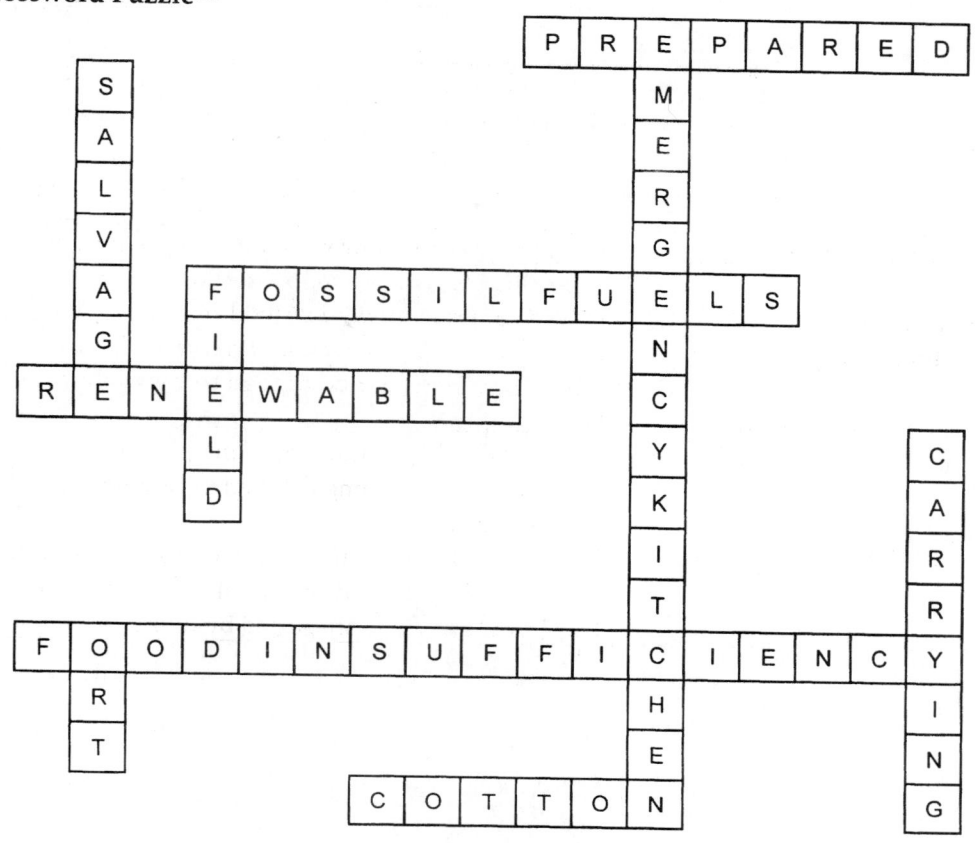

249